W9-DHM-365

TELECOMMUNICATIONS

TELECOMMUNICATIONS

THE STORY OF TELEVISION

THE LIFE OF

Philo T. Farnsworth

By GEORGE EVERSON

ARNO PRESS
A New York Times Company
New York • 1974

Reprint Edition 1974 by Arno Press Inc.

Reprinted from a copy in
The Newark Public Library

TELECOMMUNICATIONS
ISBN for complete set: 0-405-06030-0
See last pages of this volume for titles.

Manufactured in the United States of America

Library of Congress Cataloging in Publication Data

Everson, George, 1885-
The story of television.

(Telecommunications)
Reprint of the ed. published by Norton, New York.
1. Farnsworth, Philo Taylor, 1906- 2. Television. I. Title. II. Series: Telecommunications (New York, 1974-)
TK6635.F3E8 1974 621.388'092'4 [B] 74-4677
ISBN 0-405-06042-4

THE STORY OF TELEVISION

THE LIFE OF Philo T. Farnsworth

THE STORY OF TELEVISION

THE LIFE OF Philo T. Farnsworth

By GEORGE EVERSON

W · W · NORTON & COMPANY · INC · *New York*

PRINTED IN THE UNITED STATES OF AMERICA
FOR THE PUBLISHERS BY THE VAIL-BALLOU PRESS

TO

HELEN and JESS

Contents

Illustrations

Introduction

THIS BOOK is a tribute to the inventive genius of Philo T. Farnsworth, one of the greatest yet least publicized scientists of our generation. It also stands as a tribute to the American way of life, in which ingenuity and progress are encouraged by our system of free enterprise, and to the courage, vision and faith of modern pioneers of American industry such as George Everson and Jesse McCargar.

This story of Philo Farnsworth, who through perseverance and unending research rose from an obscure farm boy with an idea to a famed inventor with a discovery that is enriching our living, contains all the elements for a Horatio Alger tale. But the story of Farnsworth is true. Moreover, it didn't take place in the days of Thomas Edison, Alexander Graham Bell and other great inventors—a period when America was "growing up" and when the vast field of science was first opening to historic discoveries. This story has occurred entirely during the twentieth century; it belongs to our generation.

Farnsworth was a teen-aged youth when in 1922 he conceived his scientific ideas for an all-electronic television system—the system that provides the basis for television in use today. At the age of fifteen he astounded his high-school science teacher by explaining in logical manner, with diagrams on the schoolroom blackboard, how he planned some day to transmit and receive images over distances of many miles. The fact that such a concept could be propounded by a high-school freshman in a remote town in Idaho was amazing enough, but it is all the more astonishing when it is remembered that this youth had never been close to a research laboratory or a radio broadcasting station.

Ironically, it may be that young Farnsworth's isolation from scientific centers and his lack of knowledge as to experiments then being conducted in television were a help rather than a hindrance to him. For at that time, in 1922, television experiments going on throughout the world involved the use of scanning disks or other mechanical means for transmitting and receiving pictures. This Utah-born farm lad proposed to telecast pictures through the use of electronics, with no moving parts, and it was this concept that was eventually adopted and put into practical use. Where learned scientists in foreign countries as well as the United States had failed, Farnsworth succeeded. And he did so while the world's great scientific thinkers were saying that it could not be done.

However, the path between the conception of an idea and its fruition into reality and common use is often a long one. Farnsworth learned that as year after year went by without his invention being successfully completed. Lack of funds to carry out his objective was, as is so often the case, the big stumbling block. In fact, it was four years later, in 1926, before he was able to get any financial assistance and start building the first working model of his television system.

That assistance came from George Everson, at that time a community chest campaign organizer, who hired young Farnsworth to assist in a business survey in Salt Lake City. The two soon became close friends, and it wasn't long before Farnsworth impressed Everson with the importance of his television idea and gained the latter's assistance in forming the research laboratory that was to chart the future course of the new television science.

In this book, as told so capably by Everson, is the complete story of Farnsworth's long struggle, from his early days to his eventual success and recognition as the "father of television"; his experiences in originating and having credited to him more than 100 electronic and television patents; his constant battle to protect and retain rightful credit for his discovery; his battle against illness and financial hardships.

It has taken more than two decades of constant, painstaking research to bring television to its present state of excellence. It has been a costly project involving millions of dollars. It would have disheartened the average man to invest so much time and energy in developmental work whose completion was "just around the corner" for so many years. But the small group which blazed the television trail—nurtured it through its formative and trying years of the 1920's and 1930's—was not composed of ordinary men. They shared a conviction and foresight with which most of us are not endowed.

Philo Farnsworth often has told me that he has always felt, and still feels, that there is no problem which cannot be solved by man if he intelligently applies himself to the task long enough and diligently enough. As the pages of this book reveal, the men who backed the youthful inventor in his early work likewise believed there was no such thing as an insurmountable obstacle. George Everson and Jess McCargar, a California banker and friend of Everson, succeeded time after time, often against overwhelming odds, in raising the vast sums of money required to finance his research. They gave him their full confidence. And ten years ago they took the lead in forming the company that now bears his name. Both men have continued to help guide the company by serving as directors. The millions who will receive the benefits of this new medium of communication owe to them a debt of gratitude.

There is no person better qualified than Everson to serve as Farnsworth's biographer. He not only was the "discoverer" of Farnsworth, but he has been a loyal friend and adviser for nearly a quarter of a century. He has shared with the inventor the many disappointments, successes, and problems that have accompanied the incalculable number of hours spent in developing and perfecting electronic television.

Although anecdotes about Farnsworth have been a popular topic for writers and speakers, the full story of his historic work in television has never before been told. It is given in its entirety for the first time in these pages. It is, I believe, a particularly fitting time for it to appear in print. For television has now truly arrived, bringing with it new concepts of education and enjoyment. Television receivers are already installed in several hundred thousand American homes, and programs are being viewed daily by millions of persons throughout our nation. This amazing new science has now taken its place alongside the other great scientific discoveries that are enabling all of us to enjoy finer, fuller living.

And it all boils down to this story of a young man with an idea—in a land of liberty and opportunity—and a group of far-sighted men who were not only permitted, but encouraged, by our thriving system of democracy to help that young man make his idea work.

EDWIN M. MARTIN

(((1)))

A Boy's Purpose

PHILO T. FARNSWORTH, the son of Lewis and Serena Bastian Farnsworth, was born on a farm near Beaver City, Utah, on August 19, 1906.

Both the Farnsworth and Bastian families had been Mormon pioneers. The inventor's paternal grandfather, for whom he was named, was one of Brigham Young's lieutenants. He assisted in the erection of the Mormon Temple in Nauvoo, Illinois, which was destroyed in 1848, and thereafter followed Young on the long migration to the promised land of Utah. Having gained a reputation for endurance, good judgment, and loyalty, he was commissioned to establish the Beaver City community in the southern part of the territory. Later he became the Mormon Bishop of Beaver City, probate judge of the county, and a member of the Territorial Legislature.

The inventor's mother's forebears were of equally rugged stock. His maternal grandfather, Jacob Bastian, was a member of the Mormon colony from Denmark. At Iowa City he joined the band of Mormons and pooled his resources with them for the western journey. Jacob could not speak English, but his wife was an accomplished linguist. While he had sufficient funds to fit his own family out with a wagon train, others of the band were not so fortunate. When all of the resources were thrown together for common use, there was just enough to fit out each family with a pushcart for the difficult trip over the plains and mountains to the West. Only seventeen pounds of luggage were allowed each member, so many of Jacob's cherished household possessions brought from the old country had to be abandoned.

It was a heartbreaking and possibly a foolhardy venture. Many perished by the wayside. Jacob Bastian and his wife arrived at the promised land after 1,300 miles of hardship, but his wife died shortly afterward. He married again, his second wife being a girl of fourteen who had also come to Utah with the pushcart pioneers. They settled in Moroni, where he became the expert carpenter of the community.

The home in which the grandson of these pioneers was born was a modest farm dwelling just one step beyond the log-cabin stage. The livelihood of the family was gained from the produce of irrigated acres. When the farm work slackened, the father added to the family's income as a teamster with his sturdy draft horses.

Lewis and Serena Farnsworth were indulgent with their children and eager to give them as many educational advantages as their limited circumstances would permit. They subscribed to one or two popular technical magazines which were often the center of family interest, and which afforded subjects for discussion and speculation in the long evenings where there was no radio and where books were at a premium. There were the Horatio Alger–like accounts of inventions that stimulated the imagination and led Philo, at the age of six, to make his bold declaration of intention to become an inventor. His father encouraged this ambition by keeping up the subscriptions to the periodicals and listening interestedly to the child's imaginative flights about building things and making them work. At the age of twelve, Philo's interests had grown to such an extent that he entered a contest in one of these magazines. He won the $25 first prize. With it he bought his first long pants.

In 1920 young Philo's uncle persuaded the boy's father to join him in taking over a ranch in Idaho. A caravan was made up of livestock and farm wagons to carry the family furniture and belongings to the northern location. As they trekked slowly through the streets of Salt Lake City, the fourteen-year-old Philo, called Phil by his family, stood in the back of one of the wagons and gazed in open-eyed wonderment at the city through which they were passing. As good Mormons, they stopped in respectful curiosity to wander through the grounds of the Tabernacle and the Temple and to gaze at the beautiful monument to the sea gulls which legend credits with saving the first Mormon crops from destruction by grasshoppers.

Most eager and curious of the migrants was young Philo. Here at the center of the Mormon faith he first saw the great Tabernacle and Temple as living symbols of what the energy and industry of man can build if he has a purpose.

Leaving Salt Lake City, the farm caravan moved north over the rough and barren terrain of northern Utah and southern Idaho to the eastern section of the state, which borders Yellowstone Park. Among the household goods were a few of his greatly prized possessions—the products of his early inventive genius.

With the long journey from southern Utah to the upper Snake River country ended, Phil's father and his uncle and the family settled on Bungalow Ranch in Rigby, Idaho. It was a large hay and grain farm with two big white houses, huge granaries with automatic elevators, and all sorts of outbuildings for farm use. Phil's uncle was buying the ranch, and his father was helping in the undertaking.

Among the things which engaged the interest of young Phil was a Delco lighting system, which was used not only for light but as a source of power for many of the household and barn chores. This power plant was new to Phil, and it completely engrossed his interest. It was his first hand-to-hand contact with any type of electrical equipment. He was the only one in the family who knew how to run it successfully, and it became his pleasant duty to keep it in repair. He was the happiest when it was out of order and needed fixing. In fact, there is a suspicion that Phil often put it out of commission in order to have an opportunity to take it apart and put it together again.

Phil was constantly thinking up new ways to utilize the resources of the plant. Most of them remained in the realm of his youthful imagination. One, however, was reduced to practical use and great personal advantage. He devised a motor to harness the power of the lighting system for his mother's washing machine to relieve himself of the drudgery of operating it by hand.

At the opening of the school year Phil entered the local school to prepare for high school. On his graduation he registered at Rigby High School. Upon entry he had a long talk with the school superintendent, Justin Tolman. Tolman was one of those fine men who dedicated their lives to teaching. He was quiet-spoken and firm but profoundly understanding. Phil in his eager way wanted to encompass the essentials of the whole high-school course in one big gulp of educational assimilation. Tolman told him that he thought it would be wiser to take the first-year course as prescribed and supplement his work by outside reading. He suggested that there were a few books in the library that might be of interest to him.

Phil took Tolman's advice and registered as a freshman, but he soon found that the prescribed courses were not sufficient to keep him busy and interested. As Tolman had suggested, Phil ransacked the library for books on scientific subjects and found an electrical encyclopedia which he eagerly pored over to supplement the electrical knowledge he had obtained by reading the amateur technical magazines his father had secured for him. He talked with students in the higher grades to find out the gist of their work.

One morning Phil came to school early and approached Tolman, who was sitting at his desk busily preparing for the day's work, and asked to join the senior chemistry class, which was taught by Tolman. The request was an astonishing one, since entering the class in the middle of the term would require that he make up the first three months' work of the course. Tolman told him that he would have to wait until his senior year, that it would be folly for a freshman to think of doing the work of a senior. The boy left, and Tolman thought he had seen the last of him in the chemistry class for three years. But he was mistaken: the next day Phil appeared again and asked the privilege of just sitting in with the group. The request was granted, and Farnsworth soon proved a worthy student. His questions were so penetrating that he often stayed after school with Tolman to thrash them out. It finally resulted in his coming early and staying late and getting special tutoring from Tolman beyond the requirements of the senior-class course. In fact, his hours before and after school were so long that it brought a complaint to the school authorities from the janitor.

Tolman had the true teacher's delight in leading a young

and active mind into new fields of knowledge. The after-school tutoring, therefore, became not only hours of work, but periods of pleasant and friendly understanding out of which the teacher got as much pleasure and profit as the pupil.

It was not extraordinary, then, that Phil confided in Tolman his purpose to become an inventor and gave it as the principal reason for his eager pursuit of scientific knowledge. As a result their discussions went far beyond the chemistry textbook.

Because of the intellectual companionship he enjoyed with the teaching staff, Phil was often put in charge of the high-school study hall. One day Tolman broke in on the study hour unexpectedly and found Farnsworth giving the assembled students an exposition of the Einstein theory of relativity. Tolman remained quiet and listened as Phil unfolded the mystery of the theory with simple clarity and dramatic force. Here, as always, Phil was a good salesman. He made the scientific concepts live and glow with his own enthusiasm. Later Tolman said that it was the best discussion of the theory of relativity he had ever heard or read.

For one of his years Phil had an amazing grasp of scientific subjects. Treatises in encyclopedias gave him the same thrill that a Nick Carter story held for most boys. His curiosity on technical matters knew no bounds. Reading about one led to interest in others. It was fun sleuthing through abstract pages of scientific exposition to get at the meat of things. He had the rare gift of visualizing theoretical concepts and making them live in his imagination. The electrical encyclopedias were not to him forbidding volumes of lore to be avoided; they were mines of interesting facts and theories to be understood and utilized—they were the Arabian Nights of the world of learning.

From them he learned about the electron theory. As his

studies progressed, the little vital units that scientists had given the name of electrons became realities in Farnsworth's mind. He inquired into their properties, what influenced their behavior, and how by harnessing them men were sending and receiving audible messages and music through the air.

Again and again he returned to the device that, according to the book, changed light energy into electrical current, and its counterpart that changed a beam of invisible electrons into a glowing light. The first was the photoelectric cell; its opposite, the cathode-ray tube. The key to understanding all this was mathematics. It meant study, but study that was fascinating and fruitful.

Phil continued to enjoy astonishing his young brothers and sisters with his vivid descriptions of what he had learned of this and that in science, how he was going to make great things from his knowledge and become famous. The two new instruments were additional wonders to explain to them. In the midst of one of his eloquent expositions Phil became pensive. A conception was taking form in his active mind. He saw the photoelectric cell and the cathode-ray tube as teammates, one translating light into electrical energy, and the other changing electrical current into glowing fluorescence.

The next day he returned to his encyclopedia to study. He walked alone to give his imagination free play with the descriptions of the two tubes, to visualize more clearly what electrons might be made to do within them. Here were two instruments made to order for the transmission of pictures over great distances, whether by wire or through the air. It was a perfect field for the exercise of inventive genius, a God-given opportunity for study and development.

The idea of television itself was not new in the realm of the human mind, but Phil Farnsworth's conception of harnessing the photoelectric cell and the cathode-ray tube in a simple television system was new and original. At this time he was not aware of what had gone before in the efforts to transmit pictures. He knew that, through the telephone and radio, sound could be transmitted almost instantaneously around the world by electrical means. Why, then, he thought, with the aid of the photoelectric cell and the cathode-ray tube would it not be possible to do the same thing with visual images? Eagerly he concentrated on the problem.

He thought it through, working out each step in a careful diagram, planning how he would handle the variations of light and shade in a picture, translate them into their electrical counterparts, send the varying current through the air, and restore it in proper sequence at the receiving end in glowing light. His conception of the sending tube resulted in the dissector tube, which became the heart of the Farnsworth television camera. Essentially it was a simple but revolutionary device. This idea, produced by a youth of fifteen, was later characterized by Dr. Herbert E. Ives, of Bell Laboratories, as the most daring invention of which he had knowledge. Though it was years later that Phil named it the dissector tube, at this early date he clearly had in mind how it would function.

The scheme became an obsession with him. He must tell someone about it. He must discuss it to clarify the details in his own mind. Tolman was the logical confidant. Finally, when he felt he had the plan well worked out, he determined to reveal it to his mentor.

The study hall with the large blackboard at the front was usually vacant during the last period of the day. With the routine work for the day accomplished, Phil would repair to this room to prepare himself for the after-school session in chemistry with Tolman. This seemed to him the logical place to reveal to Tolman the details of the television scheme that had been driving him during most of his waking hours for the past months. With the sense of the dramatic which was growing to be an essential part of his personality, he planned to surprise his instructor by drawing a full outline of his scheme on the blackboard before Tolman came in.

When Tolman entered the room, the boy was not in his customary seat poring over a book, but was finishing a blueprint type of diagram that covered half of the blackboard which stretched across the front wall of the room.

Tolman stepped to the front of the study hall and sat down to watch the boy as he completed his diagram with nervous, almost fidgety strokes. When he had finished the drawing he placed chalk and eraser on the base of the blackboard, walked over to the end, and picked up a pointer. Returning to the center in front of Tolman, he announced with eager, half-suppressed excitement that this was his new television system.

Tolman asked what this had to do with chemistry. Phil then went on to explain that this new invention of his had really been occupying his thoughts for a long time and he wanted to explain it to him.

With that as a start he went through the whole diagram, giving Tolman, with glowing enthusiasm, a clear, concise explanation of how the whole television scheme would work to bring pictures of living scenes from a distance into homes, as sound radio was then beginning to be brought to the homes of America. Tolman listened with rapt interest. He had had enough experience with Phil to know that what he was saying must have some basis in science or the boy would not have worked so hard on it.

There was not time between the close of the school session and the supper hour to go into all the details of this scheme as

Phil visualized it. It took many more evenings. In these sessions, chemistry was forgotten and Phil stood before the blackboard posing the problems and solutions embracing the general system of television. It must be remembered that this plan was unfolded evening after evening in the Rigby High School by a fifteen-year-old boy in 1922, when sound radio was yet in its infancy.

Getting down to details, Phil asked Tolman why it would not be possible to focus the image to be transmitted on the photoelectric surface in the vacuum tube. If this were done under proper control, each point of the image would give off a flow of electrons representing the strength of the light focused on that spot. By this simple device could he not build up within the vacuum tube an electron image which would correspond exactly to the picture image focused on the surface?

This was indeed a startling conception as yet unknown to science. It is true that the electron image within the vacuum tube would be invisible, but nevertheless it would be an exact reproduction of all the lights and shades of the actual image in unseen electrical units.

In thinking the matter through, Farnsworth concluded that unless he prevented it the electron image would unquestionably become blurred. His knowledge of optics told him that light beams could be focused; therefore, why could he not provide a magnetic lens, or solenoid, to control the electrons and keep the unseen electron image sharply focused? This magnetic focusing was the second essential in the development of his television camera.

In the transmission of pictures, Farnsworth knew that he must break up the image and transmit it a unit at a time. In other words, the image had to be scanned in much the same manner as the eye reads a page of print. In considering the problem, the shape of the tube became important. He conceived that the tube must be a cylinder closed at each end by flat surfaces. The front of the tube would be an optically clear window through which the picture would be projected onto the photosensitive coating of the rear plate of the cell. The impact of the varying light intensities of the picture on the photoelectric surface would release an electron image. This would flow forward to the front end of the tube, where the scanning process would take place.

To accomplish the scanning Phil devised an anode finger projecting up in the tube to act as a "collector of electrons." He thought of it as a small metal cylinder about the size of a lead pencil, with a pinpoint aperture into which the electrons would flow and be carried out as varying electrical current. This would constitute the picture signal.

To control the flow of the picture elements into the aperture, means had to be provided for scanning the image across the anode aperture.

He would do this, Farnsworth told Tolman, with magnetic scanning coils, which by attraction and repulsion would oscillate the electron image back and forth in an orderly fashion over the anode slot. He realized that two sets of coils would be necessary, one acting rapidly to scan the image a line at a time in a horizontal direction, the other, at a slower pace, to move the image up by degrees as the lines were scanned. By this method the picture signal would flow out from the tube through the anode in orderly sequence.

Such was Farnsworth's original conception of a television camera. In it a picture was translated from light values into a ribbon of electrical variations which then could be handled exactly as any other electrical current. Patiently he outlined step by step each part of the transmitting mechanism so that

Tolman could see what he was driving at, as eagerly he strove for approval. Tolman understood and told the boy that it sounded reasonable.

Then came the problem of receiving the image in the home. Farnsworth's idea was to build a receiving set around the cathode-ray tube. To do this it would be necessary to devise means by which the picture signal radiating through the ether could be detected by the household television set in a manner similar to that used in sound radio. Amplifiers would be required to build up the electrical current representing the picture values and feed it into the cathode-ray tube.

The tube which Phil proposed to use as the heart of the receiving set was a pear-shaped vacuum bulb with a slender stem. A current flowing through and heating a filament in the stem gave off the electrons which formed the invisible cathode beam. Phil visualized this beam as proceeding in a straight line to bombard. the fluorescent surface lining the opposing bulb end of the tube, thus causing it to glow. This completed the cycle of translating electrical current back into light values, which in turn would reproduce the original image for viewing in the home.

To provide the lights and shades of his picture, Phil reasoned that if he put a shutter in the stem of the cathode-ray tube to control the flow of the beam it would be possible to use the varying picture current coming from the transmitter to operate the shutter, or grid. However, it would be necessary to focus the beam and oscillate it for scanning. Here again Farnsworth's inventive mind provided a magnetic lens in the form of a coil around the stem of the tube to bring the beam to pinpoint sharpness. Magnetic scanning coils similar to those at the transmitter would also be provided. Operating in exact synchronism with the pulses at the transmitter, these coils would control the beam so that it would etch an exact reproduction of the original picture on the cathode surface. Because of the persistence of vision, the varying light intensities created by the moving pencil of electrons would give the illusion of a perfect image.

This was Farnsworth's conception of modern television, and it was this that he conveyed to Tolman.

One can visualize the eager youth evening after evening sketching out drawings of the scheme as he conceived it. One can hear his mentor asking pointed questions and raising doubts here and there as the plan was unfolded. One can catch the impetuous flow of explanations in Phil's anxious bid for approval, and later the quiet discussion of the plan, a point at a time, searching for holes in it.

Many such hours of discussion must have taken place, because six years later, without any intermediate discussion or coaching, when Tolman appeared as a witness in a patent suit he sketched from memory in accurate detail the television scheme that is now the basis of the whole Farnsworth system.

Day after day, when the advanced chemistry lesson had been completed, teacher and pupil continued to discuss the all-absorbing idea of Phil's television scheme. It was something made out of whole cloth from the imagination and ingenuity of a boy's inquisitive brain. In theory they worked out the scheme, anticipating the difficulties they would meet and overcoming them in imagination when they seemed insuperable. Gradually it became the consuming interest in the boy's life. He talked it over with his parents, who had more than sympathetic interest in their son's chosen profession of invention.

Farnsworth had other schemes that he thought patentable, simpler devices than the television system. His father advanced money to file a patent application on the one that seemed most likely to bring immediate returns. The thought was that if this were successful the revenue from its exploitation could be used to further the plans for television. Phil's idea was forwarded to a Washington firm of patent attorneys advertising in an amateur's magazine. The case should not have been filed, for the idea had little commercial value. In addition, it was handled badly, so that nothing but additional expense and disappointment resulted.

Phil's interest in invention and scientific subjects did not prevent his being quite a figure in high-school activities. He was a violin player in the high-school orchestra and a leader in the social activities which centered around the school. He also learned to play the piano by the "hunt and pick" system. To this day he enjoys nothing more than to scrape away at his fiddle or improvise at the piano. His music teacher was so impressed with his ability as a violinist that she urged his father to encourage Phil to take up music as a profession, but Phil was one not easily influenced; he had a mind of his own, and a purpose. Music to him was a diversion from the concentration of thought centered on his invention.

Farnsworth's genius and his great ability to concentrate had its inception in the period spent at Rigby High School. He set about to train his mind in constructive thinking and purposefully worked out a plan of mental discipline that gave him an enormous capacity for quick comprehension and ready adaption of new ideas and information to the purpose in hand. He had a habit of waking up an hour before rising time and thinking through the problems that were puzzling him, because in the morning his mind seemed to be clearer and more readily responsive to constructive effort. This habit of mental concentration has been carried on throughout his career.

(((2)))

Frustration and Disappointment

PHIL'S FIRST YEAR in high school was one of great intellectual growth. The seminars with Tolman, the technical encyclopedia, and his regular classroom studies gave him some inkling of the vast scope of scientific knowledge. New fields were open for conquest, and he delighted in accepting the challenge. It came to an end all too soon, for at the close of the spring term it was necessary for the family to move to a part of the general ranch holdings which was outside the school district. He therefore could not continue his high-school work in Rigby without prohibitive tuition fees. The distance to and from school was also too great for the rigors of the winter months. The educational facilities within the new district were not up to the standards that Phil required.

There seemed to be no point from which he could proceed in a practical way with the ideas that were teeming in his brain. His father was keenly interested in the boy's future and sympathetic toward his television inventions. He encouraged him to persevere, but there was no money for further patent attorneys' fees, or for the filing of patents. Worse still, opportunity for further study seemed to be denied. It was a hopeless situation.

One of his half-brothers was then working in the railroad shops at Glen's Ferry, Idaho. Finding himself blocked in his efforts to further his education, Phil packed up such books as he could lay his hands on and went to Glen's Ferry, where he got a job as an electrician with the Oregon Short Line. One of his duties was to see that the locomotive headlights were focusing properly before the train left the station. He speaks feelingly even to this day regarding the frightening experience of crawling out over the boiler to adjust the headlight when the engine was in motion as it switched around the yards. For a sensitive, frail youth this was an ordeal, particularly if there was a blinding snow or rain storm.

Phil was the oldest of a family of five children, having two sisters and two brothers. Phil's disappointment in the educational facilities available, and the need for better school advantages for the other members of the family, led the father and mother to take counsel concerning the future of their family. After careful consideration they determined to give up the ranch and move to Provo, Utah, the seat of Brigham Young University. Phil followed the family to Provo and entered the high school there in 1923 in preparation for entrance to the university. There is a close relationship between the high school and the university, so from the time he entered Provo High School he had the run of the university research laboratory.

Brigham Young University is sponsored by the Mormon Church. For a school of its size, its faculty is exceptionally well qualified, particularly in the field of science. Some outstanding men have been developed by this small but excellent university. It has preserved the originality of thought, the attention to essentials, and the development of initiative among its students that have made the Mormons so successful in conquering a pioneering country under most adverse conditions.

While it was necessary for young Farnsworth to complete his high-school course before formally entering the university, he was able to pursue studies in higher mathematics by working with his cousin, Arthur Crawford, who was enrolled as a regular student. After a year of work, Phil finished the high-school course.

He had been unable to make any progress in exploiting his television ideas during this time, however. He chafed at the limitations that necessity placed around him. When the school year ended he was not successful in finding employment for the summer. In the desperation of frustrated youth he and one of his schoolmates were attracted by a recruiting campaign put on by the Navy in Provo. Here he saw a way to prepare himself for enrollment in Annapolis.

The Bay of San Diego, the sea, and Navy life were all new and exciting to Phil in the beginning. However, as time went on, the routine became irksome and uncongenial to the young inventor, whose main interest was in science. The more he learned what might be ahead of him, if assigned to the Naval Academy, the less it seemed suited to fit him for the career he wanted.

While he enjoyed the association of the boys at the station, his best friend was the chaplain, who soon recognized Phil as a youth of exceptional ability. With him he had long discussions on how he could best be fitted for a career of scientific invention. To the chaplain it seemed that training Phil for the Navy was something like putting a spirited race horse to draft-horse duty. Therefore, in spite of his loyalty to the Navy, the chaplain had to counsel Phil against training at the Academy and gave consideration to taking steps to have him released from the Navy to return to Brigham Young University.

Then a crushing blow fell upon the Farnsworth family. Phil's father contracted pneumonia and died shortly before Christmas of 1924. Phil's mother sorely needed the help of her oldest son, and through the efforts of the naval chaplain and a family friend, his release from the Navy was secured.

Upon his return to Provo, Phil and his mother worked out plans to keep the family together and provided for. He got part-time work to contribute to the family income, and by careful planning he was able to enter the university as a special student. Here he spent every spare moment in scientific studies. It was a great opportunity, and he made the most of it.

With the fear, common with inventors, that the idea might be stolen from him, Phil was very careful whom he told about his television scheme. He did feel, however, that he should discuss it with someone who might give him additional help. He talked with his cousin, Crawford, who advised him to discuss the matter with some of the faculty members of the university. This was one of the first things he did after enrollment. The faculty members were most sympathetic and helpful. They permitted Phil to take the mathematics and electronics courses that would be helpful in perfecting his invention. They also gave him the run of the laboratories and allowed him the use of the meager facilities. This was heaven for Phil.

In his classwork he specialized in mathematics, since it seemed to be the most necessary subject for him to pursue in order to develop a basic understanding of the physics and electronics involved in his venture. He was a very apt mathematics student and was the delight of the faculty who acted as his mentors. Two of the professors in particular considered him as their special protégé. They took interest in feeding his inquiring mind and opening paths into new fields of knowledge.

It was here that Phil first came in contact with the Bell Laboratories journals and their content of abstruse papers on electronics. Here he first learned of Hamilton, the great scientist in the theory of light. One test in the laboratory proved that a beam of electrons could be deflected magnetically. This was one of the fundamentals of his scheme. The proof that a beam could be deflected gave Phil added confidence in his plans.

As they followed Phil's ideas on television, the faculty members became more and more interested. As he went along in his studies, they led him into new scientific works and textbooks bearing on the subject with which he was wrestling. In the laboratory they watched with interest as he made use of the facilities.

Here in Provo he conceived other ideas for simple inventions which he hoped might be exploited to finance the television development. One was a vernier dial. Another was a clever type of ignition lock. None of them was successfully commercialized, and there seems to be a record of only one of them being patented.

The television scheme became an absorbing interest to the faculty members to whom Phil gave his confidence. There were many sessions of discussion as to how the thing could be exploited, but nothing practical came out of the talks.

As in Rigby High School, Phil was active in the social doings of Brigham Young University and of Provo in general. He played the violin in the school orchestra and was first violinist in the Chamber Music Orchestra. It was here that he met Elma Gardner, an attractive girl of his own age, who had wit and intelligence to match his own. They were leaders in a group that enjoyed music, dancing, hiking, and the general run of social activities in the Mormon Stake to which they belonged. A Mormon Stake is a local unit of the Mormon organization which, as Vice-President Thomas Marshall once observed to Mormon President Heber Grant, corresponds almost identically to the ward in the Tammany organization in New York City. Life at this time was interesting and intense for young Farnsworth, divided as it was between his inventive activity, his school work, his violin, and the normal fun-making of youth.

The school term, which began in January, seemed too short. He had enjoyed this first experience of university life, and he wanted to continue. To assure his next year's education he lined up a janitor's job for the following school session.

At the end of the term Phil joined his brother in a lumbering operation in Payson Canyon. He worked hard all summer and returned to Provo two days before the opening of the fall term in 1925 to find that his janitor's job was not available. In spite of this setback, he enrolled in the university and did such odd jobs as he could find along with helping his mother in her effort to keep the family together. This proved not to meet their needs, and he finally realized that his opportunity for further school work was blasted.

This was in the comparatively early days of radio when there were few technicians qualified to install and service radio sets. Phil's technical studies gave him some knowledge of the subject. Talking it over with his mother, he decided that the new field should offer opportunities for work and advancement. Practical acquaintance with the technical problems of radio also might afford chances for furthering his television scheme. With these considerations in mind, he went to Salt Lake City early in 1926 to get a job in radio. It was not easy to find what he wanted. Finally he set himself up independently in a radio service shop.

Cliff Gardner, brother of Elma, had just finished high school, and he joined Phil in the Salt Lake venture. It was a hand-to-mouth existence. In spite of the demand for radio service and the excellent work done by Farnsworth, he found himself always in debt to the man who provided the meager headquarters for his business operations.

Since the art was new, the installation of a radio set, to give good reception and dependable service, was a real job. Phil did the work well and gave general satisfaction to his customers. However, his efforts and those of Gardner hardly sufficed to keep them in food and shelter. They lived in a furnished apartment and cooked their own meals. The situation became so desperate that it was obvious that this was not the way for him and Gardner to make their livelihood.

In the spring of 1926, in an effort to better his condition, Phil registered with the employment agency at the University of Utah. He was not discouraged, but he was faced with the problems of immediate necessity and of giving consideration to the future comforts of the widowed mother and younger brothers and sisters.

(((3)))

Promise of Financial Help

IN THIS SAME spring, Leslie Gorrell and I were driving from San Francisco by way of Los Angeles and the Mojave Desert to spend two months in organizing and raising the first year's funds for a community chest in Salt Lake City.

While we were crossing the desert a bearing burned out, and my car had to be abandoned at St. George, Utah, forty miles from the nearest railroad. We took a bus from there to Beaver City and proceeded to our destination. The car was to be brought on by a mechanic when the repairs were finished.

Upon arrival in Salt Lake City we got the preparations under way for the campaign. As a detail of organization it was necessary to make a business survey. Following our usual custom, university students were sought for the work.

Philo T. Farnsworth was one of those who came to the Community Chest headquarters from the employment office of the University of Utah. He and several others were engaged. Fortunately, Farnsworth was placed in charge of the survey crew of six students.

Farnsworth found out that the Chest headquarters needed

an office boy, with the result that his friend, Cliff Gardner, applied for and received the job.

A room was set aside for the survey work, maps were drawn of the business district and plans laid out for a careful door-to-door check in order that all of the business houses and their employees should be tabulated for the prospect list of the fund-raising campaign. It was necessary that young Farnsworth have some clerical assistance. In discussing it Phil volunteered, "Mr. Everson, I know just the right person for this job."

"Bring her in," I said, "and let me talk to her."

"She's down in Provo," Phil replied. "She can be up here in a day or so if you are sure there is a job."

"Just the right person" was Miss Elma Gardner, who came up from Provo two days later. She turned out to be most capable, and, aside from his personal interest, Phil's selection was properly in the interest of efficiency and good office management.

Shortly after Phil came on the job, word came from St. George that my car had been brought as far north as Beaver City by the mechanic. There the bearing had again burned out, and the car was being held at a garage for further instructions. I telephoned the garage again to make the necessary repairs and to notify me when it was finished. I told them I would then send someone for it, as I needed it badly in the campaign work.

The local young men on the campaign staff learned of the situation. Several of them offered to fetch it. Young Farnsworth particularly wanted to go, and with his characteristic purposefulness, he got the job. Two days passed with no word from him. Finally at the end of the third day, when I was almost distracted with worry about him and the car, a long-distance telephone call from Provo came in. It was Phil.

"Where are you?" I asked with some impatience.

"I'm in Provo," he replied. "The bearing burned out again. I couldn't get it fixed, so I pulled out the piston and have limped in to here on five cylinders. I'll be in late tonight."

I told him I wouldn't wait up for him but would see him in the morning.

The next morning Farnsworth impressed me as knowing more about the car than the man who made it, so I asked him to go along with me and give instructions to the mechanic for its repair. Needless to say, the car was successfully repaired and gave no further trouble. This event brought young Farnsworth and me closer together and opened the way to a long series of conversations about his hopes and dreams of that then-fantastic idea—television.

A fund-raising campaign at best is strenuous and nerve-racking. If the campaign effort is to be successful, everything must function according to a prearranged schedule. Phil didn't quite understand this necessity. I had given him the responsibility of getting out a mailing on a certain day. When closing time came it was hardly begun. With some impatience I asked him why he hadn't followed instructions.

"Oh," he said, "we'll get it out tomorrow."

"But that won't do, Phil," I replied, "it must go out tonight. The only thing to do is to get a hurried supper, then we'll all pitch in and get it out."

Phil was nettled about it, and the others, seeing his vexation, came back after supper. The job was done by nine-thirty. The girls went home, and then Farnsworth, Gorrell, Gardner, and I leaned back in the chairs around the big mailing table and started a sort of bull session.

I asked Phil if he planned to go on to school. "No," he said, "I can't afford it. I've been trying to find a way to finance an invention of mine but it's pretty tough. In fact, I'm so discouraged that I think I'll write up my ideas for *Popular Science.* I imagine I could get a hundred dollars if I worked it right."

"What is your idea?" asked Gorrell.

"It's a television system."

"A television system! What's that?" I asked.

"Oh, it's a way of sending pictures through the air the same as we do sound," said Phil.

"Where did you get that idea?" I inquired.

"I thought of it when I was in high school at Rigby, Idaho," Phil went on. "Then when we moved to Provo and I went to Brigham Young University I told a couple of the professors about it. They encouraged me and let me try out some things in the lab to prove it would work."

We talked a little further about it, but neither Gorrell nor I showed enough interest to encourage Phil to a very detailed explanation.

As we were about to break up and go home he seemed very uneasy at having said so much. He cautioned us that he had not disclosed this idea to anyone outside of his own family, one of his high-school instructors, and his professors at Brigham Young University, and he urged us to keep his confidences. Gorrell and I regarded his story at the moment as little more than the interesting daydream of an ambitious youngster. Later he discussed it at greater length with Gorrell. Gorrell was so impressed that he urged me to inquire further into the matter.

"This television scheme of Phil's has merit," Gorrell told me. "You ought to talk to him about it. I know if it's sound you can find the money to promote it."

"It is interesting," I said, "and just fantastic enough to be real. Let's ask him to dinner tomorrow night and then take him up to our apartment and really find out what it's all about."

As a result, the three of us went to dinner the next evening for an extended discussion of Farnsworth's ideas. I looked forward to the meeting with only casual concern, but it developed into one of the most interesting evenings I ever spent. When the discussion began I hardly knew what the word "television" meant. I had read somewhere that a man by the name of Baird was doing some experimental work in the transmission of pictures and that this new art was called "television." Otherwise it seemed so far removed from reality I hadn't given it any thought.

Young Farnsworth at this time looked much older than his nineteen years. He was of moderate height and slight build and gave the impression of being undernourished. His skin lacked the glow of health that is typical of boys his age. There was a nervous tension about him that was probably the result of financial worry and frustration in making headway in his scientific pursuits. Around the Community Chest office he had the appearance of a clerk too closely confined to his work.

As the discussion started, Farnsworth's personality seemed to change. His eyes, always pleasant, began burning with eagerness and conviction; his speech, which usually was halting, became fluent to the point of eloquence as he described with the fire of earnestness this scheme that had occupied his mind for the last four years. He became a supersalesman, inspiring his listeners with an ever-increasing interest in what he was saying. Once he had outlined his general scheme, we considered the practical aspects of the situation. I thought that if Farnsworth, with little more than a high-school education and a background of pioneer communities, had conceived this idea of electronic television, surely the great laboratories such as General Electric and Bell unquestionably must have hit upon the same scheme and were probably developing it in secret and had it well along to completion.

I asked Phil, "Isn't it likely that General Electric or Bell Laboratories have accomplished all you propose, and probably have it nearly ready for use?"

Young Farnsworth accepted this statement as a challenge, and launched into an exposition of what was going on in the world of experimental television work. His statement was detailed and accurate, as we found out later.

Farnsworth said that there were four experimenters in television at this time—Bell Laboratories under Dr. Ives, General Electric Laboratories under Dr. Alexanderson, and the laboratories of Baird of London and Jenkins of Baltimore. He went on to say that Baird had already transmitted rough but recognizable images.

"They are all barking up the wrong tree," Farnsworth said. "All these men are trying to transmit pictures by mechanical means. This will never do. The speeds required for scanning an image to produce a good picture are so great that there must be no moving parts. The scanning-disk system they are all experimenting with can't possibly produce commercially acceptable results."

I asked him, "How do you propose to get rid of all the mechanical parts the others are using?"

"My system is entirely electrical," Phil answered. "The necessary speeds will be achieved by manipulating the velocity of light and of electrons. If mechanical parts are used, the results will be crude and blurred."

"How do you propose to harness your completely electrical system?" I asked. "What instruments will you use?"

"I propose to do it all by manipulating electrons within vacuum tubes," Phil said.

Then in enthusiastic and convincing words he went on to explain the use of the cathode-ray tube, the photoelectric cell, and the magnetic manipulation of electron beams. It was all so highly abstruse, yet told with such conviction, that although I was in no position to evaluate the merits of the invention, I was tremendously impressed with the amazing knowledge of the youth and his certainty that he could accomplish what he had proposed.

As the evening passed I probed more deeply to gauge the authenticity of Farnsworth's knowledge and to get a basis of judgment as to whether this was a visionary dreamer with a fantastic scheme, or whether here was truly a genius charting the plans for a new and revolutionary invention.

"I should think the telephone company would have developed this idea," I commented. "The devices you propose to use would seem right down their alley. Are you sure they haven't done it?"

"Oh, I'm sure they haven't," Phil replied with emphasis. "At B.Y.U. (Brigham Young University) there is a file of the Bell Laboratories journal. I've gone through most of them. I'm sure they haven't anything but the scanning-disk system, or it would have been mentioned in the journals."

"Well, then," I continued, "what about General Electric? Are you sure they haven't something wholly electrical? Are you certain that someone hasn't broken away from the scanning-disk idea?"

Phil answered, "Dr. Alexanderson, who is in charge of G.E.'s television work, is definitely using the scanning disk—a big one. He considers it an interesting gadget but not practical. I'm sure they have nothing else."

Every answer he gave to these and many similar questions emphasized Phil's remarkable knowledge of the electronic art. His easy discussion of the technical aspects of his proposed system disclosed a grasp of mathematics that was phenomenal considering his lack of formal training. During the conversation Farnsworth marshaled an astonishing array of authorities to his defense and spoke with familiarity regarding scientific works of which the average college science student was hardly aware.

I knew nothing about the procedure for patent protection. Farnsworth had a general notion, and, like most lone inventors, he was extremely fearful of his ideas being stolen.

"The idea hasn't been patented," he said, "because I haven't had the money. Every time I pick up a scientific or amateur journal I'm afraid I'll see that someone has turned up with the same ideas I have."

He continued moodily, "There should be a foundation to which inventors could submit ideas and get backing for worthwhile inventions. I believe such a foundation would make money."

"Actually," I asked, "have you anything to patent yet? Don't you have to make a thing work before you can get a patent on it?"

"Yes," said Phil, "there is something in patent regulations regarding reduction to practice, as they call it."

"What do you think it would cost to build up a first model of your scheme?"

"I don't know," Phil replied, "but it shouldn't cost too much."

"Well, think it over," I concluded, "maybe if the cost isn't too great we can find the money to do it."

The prospect of getting someone actively interested in his venture renewed the young inventor's enthusiasm. His mind led him on into eager expositions of how it could be developed. His interest was not solely scientific, for he saw in the practical exploitation of television the open sesame to great wealth with its attendant power to pursue research in other fields. It was long past midnight when the session broke up.

A community chest campaign office is always a busy place, and as a campaign progresses the activity increases. There was little time for Gorrell and me to talk at any length with Farnsworth in the following days, yet every time there was opportunity for the three of us to get together we earnestly discussed ways and means of promoting the idea. None of us had ever had any experience in research development or the promotion of an invention. None had ever been inside a well-organized research laboratory. While Farnsworth had filed a couple of patent applications through correspondence with Washington patent attorneys who advertise in amateur mechanical magazines, he had never had first-hand discussion with a properly qualified patent attorney. Neither Gorrell nor I had to our knowledge met a patent attorney, so the three of us were about as ill-prepared for the venture under consideration as anyone possibly could be.

Yet through all the conferences snatched here and there during the busy campaign weeks there developed a determination to do something about promoting Farnsworth's idea. Phil, with superb salesmanship, snatched every opportunity to give us further details of his plans. Any scrap of paper would suffice for him to draw out scratchy designs of some part of his system of television.

Finally one day in a lull after a particularly enthusiastic campaign meeting I asked Phil, "Just how much do you think it would take to prove out your system?"

"It's pretty hard to say," Phil replied, "but I should think five thousand dollars would be enough."

"Well," I said, "your guess is as good as any. I surely have no idea of what is involved. I have about six thousand dollars in a special account in San Francisco. I've accumulated it with the idea that I'd take a long-shot chance on something, hoping to make a killing. This is about as wild a gamble as I can imagine. I'll put that up to work this thing out. If I win, it will be fine, but if we lose I won't squawk."

Later in the week the three of us sat down around a table and reached an agreement whereby Farnsworth was to give his whole time to the development of his idea. Farnsworth was to have one-half of the partnership and Gorrell and I were to share the other half. The partnership was to be known as Everson, Farnsworth and Gorrell—the arrangement being alphabetical. For his privilege of riding along, Gorrell was to pay me out for his share if we lost. Gorrell drew up a form of contract which was later checked by an attorney in San Francisco.

I pointed out to Gorrell that I had no knowledge that would warrant passing judgment on the merits of the Farnsworth proposition, but that I was willing to take a chance and gamble five or six thousand dollars on the apparent genius, integrity, intelligence, and industry of the boy. Surely my complete lack of engineering training ill fitted me to pass any critical judgment on what he was trying to accomplish.

Gorrell had graduated from Stanford in mining engineering and was more readily able to follow the explanations given by Farnsworth, but he had no more background in electronics than I had, and consequently he was hardly better prepared to pass on the technical aspects of the scheme.

Fundamentally it was faith in the ability of the boy inventor that brought about the partnership. In addition, Phil knew exactly what he wanted to do and outlined specific plans for its accomplishment. Unquestionably his purposefulness and modest self-assurance added much to my confidence in him.

Since I was to put up the money, it was I who had to make the decisions.

Farnsworth's work on the business survey was completed about two weeks before the close of the community chest campaign. As we discussed the practical aspects of the partnership agreement, it developed that Farnsworth wished to be in southern California in order to have access to the California Institute of Technology. This fitted in admirably with my plans, as my next campaign was to be in southern California.

Then a new complication arose. Farnsworth disclosed that before he left Salt Lake City he wanted to get married. He and Elma Gardner had been engaged for some time, and he wanted to take her with him to California. This contingency had not been mentioned before in the discussions of the agreement, but Farnsworth was very insistent about it, as he was about anything he wanted to do. I raised no serious objections, since I felt it would be better for Phil to have his wife with him in California than to divide his interest between Los Angeles and Provo. Also, I had high regard for Miss Gardner and felt that she would be a helpful addition to the partnership.

"Where do you plan to be married?" I asked.

"In Provo, of course, where our families are," Phil replied.

"I suppose you have everything arranged."

Phil looked up rather sheepishly. "Yes, I have, though my mother and Elma's family aren't very keen about it."

"Well," I said, "there's not much I can do about it, since it's all settled. How are you going to get down to Provo, and how long do you plan to stay?"

"We'll go down by train or bus, have the wedding, and then come right back here to get the late night train for Los Angeles."

"Since you did such a good job getting my car up from Beaver

City, why don't you and Elma drive down to Provo in it and do things in style?"

Phil was delighted with my offer of the car and went over to Elma's desk to apprise her of this pleasant turn of events. Consequently it was arranged that Phil was to be married the following week and that immediately thereafter he and his bride were to leave for southern California. On May 27 they drove to Provo as planned and were married before a Mormon bishop.

Both Phil's mother and Miss Gardner's family had strong misgivings about the future of the young couple. Phil has since confided that he too had misgivings, but was determined to take the step.

As a part of the agreement, Farnsworth was to receive $150 a month for living expenses. Also he was to have sufficient funds for the trip to Los Angeles and to get settled in an apartment there.

On the evening following the wedding Phil and Elma drove back to Salt Lake City to return my car and to say good-by before taking the train to Los Angeles.

About midnight I was awakened by a frantic knocking on my apartment door. "Who's there?" I mumbled sleepily.

"It's me—Phil!" came the anxious reply.

I stumbled to the door and let him in. Phil seemed much out of breath and worried. "What's the matter?" I asked.

"I don't think I've got enough money to go on," Phil blurted. "I guess I figured too close."

After a brief explanation on Phil's part, I wrote out a check for the additional amount and gave him the address of my business partner, Lynn D. Mowat, who was then conducting a financial campaign in Los Angeles, telling Farnsworth that he could arrange to have the check cashed in Los Angeles by Mr.

Mowat. I again wished Farnsworth luck and sent him on his way. Gorrell and I remained in Salt Lake City to complete the campaign.

Upon arrival in Los Angeles, Farnsworth and his wife went to Santa Monica for a week's honeymoon, and then visited Lynn Mowat and got his advice about where to find living quarters.

When Gorrell and I reached Los Angeles two weeks later, we found Phil and Elma well established in a four-room furnished apartment at 1339 N. New Hampshire Street, Hollywood. The apartment was one of two in the building. There was a garage in the rear.

Phil had already appropriated the dining room as his research laboratory and had purchased considerable apparatus with which to begin his experimental work. He had established a motor generator in the garage. Quite a lot of electrical equipment was set up in a closet leading off the dining room.

On the day Gorrell and I arrived, Farnsworth was greatly concerned about getting some glass blowing done so that he could build the first model of his proposed television transmitter tube. Having some free time to spend before going to work on the new campaign assignment, Gorrell and I set to work to be as helpful as possible in furthering Phil's plans. We spent several days with him in going around to the various electrical supply houses and other sources, securing needed materials and equipment.

The question of finding a glass blower stumped us completely for some time. Finally a scientific glass blower was found in one of the downtown office buildings.

"This is what I want," said Phil, as he sketched a rough picture of the tube he had in mind. "I want the inside of one end of this tube coated with a photoelectric material. It should be very sensitive. Then I want this collector at the other end with a lead wire out from it."

"That's a tough assignment," said the glass blower, "but I'll see what I can do."

We returned three days later to pick it up. The glass blower hadn't got quite the flat end on the tube that Phil wanted, but Phil thought it would do.

This was to be the first electronic television transmitter tube ever built. When completed, it was a strangely shaped vacuum cell, bulbous in formation and with one end considerably larger than the other. It was coated at one end with the photoelectric surface, and there was an electrical lead out of each end. With sufficient imagination one could recognize it as the early progenitor of the present image dissector tube, which is now the heart of the Farnsworth television camera.

As we waited in that shop for the precious tube to be carefully wrapped in cotton batting, Phil said, "Now we must get some copper wire for the focusing and deflecting coils; I've figured out what size copper wire we must use. We've also got to find some instrument for coil winding."

By great good luck the glass blower directed us to an electrical shop where we not only got the wire we wanted but found a manually operated apparatus for winding coils.

"Now we must have a counter on this to keep track of the windings," Phil said.

The clerk rummaged around and found one. Then there was shellac to be procured and heavy paper strips to separate the layers of windings.

It was early June in Los Angeles. The weather was perfect, so we set up the coil-winding operations in the back yard in front of the garage.

"Can't I wind these coils?" I asked Phil.

"Sure, but it's a dirty, sticky job," he replied.

"I don't mind that; let's get started," I said.

So the operations were getting under way, and I, with my own hands, wound the first focusing and deflecting coils for the wholly electronic television system.

It was, as Phil prophesied, a dirty, sticky job, and I, with my awkward, amateur ways of handling the wire and shellac, got all stuck up with the stuff. And I must say that the coils, like the first dissector tubes, were crudely made. For several years afterward they lay around the laboratory as evidence of my awkwardness and lack of technical skill. I don't know what ever became of them, but I surmise that Phil's pride got the better of his historic sense and he threw them in the trash can.

To one of my background, who had had no experience in electronics, the search for the elements Phil wanted in setting up his miniature research laboratory was most engrossing. We found nichrome wire at the Roebling sales branch, we got copper wire of varying diameters, we searched a lapidary shop for a crystal Phil wanted in an experiment requiring polarized light. We bought radio tubes, resistors, and transformers where we could find them to meet our needs.

As we went from store to store in the Los Angeles area, sometimes driving miles from one point to another, the subtle humor and imaginative quality of Phil's original thinking cropped out. There was a constant flow of conversation about the possibilities of the television system and about other developments in electronics.

On one of these excursions he confided to me that he believed that thought was a manifestation of electricity and that if we had electrical recording instruments of sufficient sensitivity, an accurate record of human thought could be made. He went on to visualize how whole libraries would be electrically recorded; young people would be put to sleep by some drowsing process and the records turned on; they would then be given a liberal education in the course of a week or so of sound slumber, during which facts would be recorded in the subconscious mind for use in the art of living. It has since been proved by men of medical science that thought is an electrical manifestation and that sight is recorded in the brain by the photoelectric process. It still remains for medical science to carry out the fanciful Farnsworth conception of a liberal education by electrical recording.

The work in Los Angeles was begun in May and carried on through the summer. During this time Gorrell and I spent as much time as possible at the apartment laboratory. During the day the curtains in the dining and living rooms were drawn in order that Farnsworth could work under controlled light. From time to time my roadster would drive up with large bundles of material to be carried into the house. These strange activities excited the curiosity of the neighbors.

As the experimental work got under way and the motor generator in the garage was put into operation, the neighbors had strange disturbances in their radio reception. All of this stimulated interest and engendered suspicions as to what was going on in the apartment at Number 1339. This was during the prohibition era. When people became suspicious of anything unusual going on in a neighborhood, their first thought was that it might be the operation of a still. It was not surprising, therefore, that one noon when both Gorrell and I were at the Farnsworth apartment and all were at lunch, the front and rear doorbells of the apartment rang simultaneously. Phil went to the front door and Elma to the rear door. To their surprise, each faced a burly policeman.

The policemen were polite but firm. They stated that they

wanted to search the house because there had been a report from a neighbor that a still was probably being operated on the premises. The house was duly ransacked. Nothing of an alcoholic nature was found, but the police were greatly impressed and seemed to feel that maybe they had uncovered something more sinister than an alcohol plant. They guardedly asked what all the activity was, pointing to the assorted experimental apparatus around the dining room. When Phil told them he was working on a television system, one of the policemen said, "Well, I'll be darned!" Then they both left mumbling their mystification.

The experimental work was not without its minor tragedies. In the course of setting up one of the tests I had bought $54 worth of tubes, which were put in a chassis in the closet. The motor generator was turned on and all stood around expectantly, hoping to see the results of a beam of electrons deflected by magnetic coils. Farnsworth had not guarded against the overloading of the line due to the surge as the motor generator started. The result was that when it did start, more power was turned into the tubes than they could handle and the whole batch were burned out in a split second.

As the summer wore on it became apparent that Phil's prediction regarding his ability to get at least a semblance of a picture began to fade. At best it became obvious that the only thing that could be accomplished with the funds and time available was to prove out one or two of the major principles involved.

There was a time limit to these preliminary activities because money was running short. Furthermore, by the first of September I was scheduled to be in El Paso, Texas, on a campaign contract. Therefore, late in July, after a careful discussion of future plans, it was determined that Farnsworth was to write up a complete description of his proposed television system with schematic drawings to accompany it. A stenographer was engaged, and Leslie Gorrell and Elma Farnsworth set to work preparing the finished drawings of the system from Phil's rough sketches.

Elma, or "Pem," as Phil called her, did some of the most difficult of the drawings. They were excellent and reflected her keen and active interest in Phil's work. She made it a point to carry on her mathematical studies in order to understand more fully what he was trying to do. In these early days, and in the years to follow, Pem Farnsworth equipped herself to follow the fundamental principles underlying her husband's television system. By tutoring with Phil and by independent study she maintained an intelligent grasp of the invention on which he was working.

In the preparation of the outline of his plans Farnsworth again reflected his intellectual abilities. Phil's English style was clear and concise. His scheme was presented in logical and orderly sequence. After several days of intensive work a brief embracing the entire Farnsworth television system was finished and Elma Farnsworth's and Leslie Gorrell's drawings were incorporated. It was understood that when this was finished we would make an effort to enlist the support of adequate financial interests.

With the completion of the memorandum the next necessary step seemed to be to get as much patent protection as possible and to secure some authoritative judgment as to the merits of the proposed television system. Leslie Gorrell had some college friends who were attorneys in Los Angeles. I asked him to get in touch and secure from them the name of a reliable patent attorney in southern California. The firm of Lyon & Lyon was recommended. Leonard Lyon, the senior partner, had been a lecturer on patent law at Stanford University, and his brother, Richard, was a graduate of the Massachusetts Institute of Technology.

I called on Leonard Lyon and told him the story of the Farnsworth plan.

His reaction was, "If you have what you think you've got, you have the world by the tail; but if you haven't got it, the sooner you find it out the better, because you can waste a lot of money on a scheme of this kind.

"We have arrangements whereby we can call on the California Institute of Technology for technical advice and consultation. You bring your young genius in here, and my brother Richard will bring in some qualified person from Cal Tech to join with him in passing judgment on the merits of what this young fellow has."

Arrangements were made for an appointment the following Wednesday. At two o'clock in the afternoon the conference was held with Leonard Lyon, Richard Lyon, Dr. Mott Smith of Cal Tech, Farnsworth, Gorrell, and myself present.

Farnsworth was hesitant at first but finally got into his stride and with great clarity and sincerity outlined the scheme he had in mind. As the conference progressed, it became apparent to me that Farnsworth knew more about the subject in hand than either of the technical men, and this was no reflection on their scientific training or abilities. Farnsworth completely overwhelmed them with the brilliance and originality of his conception. During the conference Richard Lyon often got up from his chair and walked the floor, pounding his hands together behind his back and exclaiming, "This is a monstrous idea—a monstrous idea!" The conference carried on until six o'clock in the evening.

At the end I said, "Gentlemen, if we are to go any further

with Farnsworth's idea it is necessary that we approach some people whom I respect, and whose good opinion I wish to retain, and ask them for financial help. I don't wish to do so unless I am completely sure of this proposition in my own mind. Therefore, I want to ask you three questions. First, is this thing scientifically sound?"

Dr. Smith said, almost bemusedly, "Yes."

"Second, is it original?" I asked.

Dr. Smith replied, "I am pretty well acquainted with the literature on the subject of electronic developments, and I know of no research that is being carried on along similar lines."

Richard Lyon said that he knew of no patents along the lines of Farnsworth's scheme, but both stated that naturally they could not know of all the developments being carried on privately or in secret by individual investigators or by the large laboratories.

Leonard Lyon stated that a patent search would be advisable as a measure of safety, and I authorized him to send Farnsworth's memorandum to his firm's Washington correspondent for a patent search before filing a patent application.

My third question was, "Is this thing feasible—can it be worked out to make a practical operating unit?"

Richard Lyon replied, "You will have great difficulty in doing it, but we see no insuperable obstacles at this time."

Then I turned to Dr. Smith and explained that I was not satisfied with the judgment of a four-hour conference and asked him to take a copy of the memorandum home to his Pasadena laboratory and give it such thought and attention as he could during the following week. I stated that at the end of the time Farnsworth and I would drive over to Pasadena and see if he was still of the same mind. Dr. Smith agreed to do this. Then I asked him what his fee was for the conference.

He named his fee and said, "I'm afraid I will have to add to that the amount of a fine for parking overtime, because I left my car on the street and came up here feeling sure I could throw this scheme into the discard in a half hour."

A week later Phil and I visited Dr. Smith at the Cal Tech laboratories. We found him more enthusiastic about the idea than he had been at the first conference.

In Phil's memorandum covering his system he had an alternate scheme of scanning in which he proposed the use of a quartz crystal. In our experiments with it we had called it the "magic crystal." In the discussion at Pasadena Dr. Smith told Phil that the mathematical background of the proposed operation of the crystal was not quite clear. Phil said, "That dates back to Hamilton's work in Dublin in 1886. Have you a textbook on light?"

Dr. Smith reached for one in the file of reference books on the back of his desk.

"Turn to the chapter on coaxial crystals," Phil said.

Dr. Smith found the chapter and shoved the book over to Phil, who thumbed through the chapter to where there were two pages of mathematical formulae. Pointing to an equation in the middle of the right-hand page, Phil said, "There it is."

Dr. Smith smiled and said, "You're right; I should have seen it."

During the following week much time was given to the discussion of ways and means of getting additional financing. Farnsworth was sure that a year's time and $12,000 in cash was all that would be required to enable him to produce a satisfactory television picture. Later developments proved Farnsworth to be the typical optimistic inventor, because it actually took more than $1,000,000 in money, and thirteen years in time, before his invention was ready for commercialization. It is doubtful that either Farnsworth or I would have had the courage to undertake the venture had we known what was before us. Both were equally ignorant of the many problems which lay between us and the final accomplishment of our ends.

In discussing future plans at this time Farnsworth felt confident that if he had available $1,000 a month for a period of one year he could produce results that would enable the group to secure recognition of his invention. I was aware that $6,000 had been spent in the short span of three months. I had also observed that Phil's concentration on his inventions made it difficult, if not impossible, for him to realize how quickly money could be spent on an undertaking of this kind.

While I had no knowledge or experience to go by, I instinctively felt that it would be better business to talk in terms of twice that amount, though Phil in his eagerness and assurance visualized the accomplishment of the whole project in a few months' time. I wanted to have sufficient funds available so that the work would not be handicapped by constant worry over lack of money. It was finally agreed that an effort should be made to secure backing up to $25,000 to underwrite the research work. It was left to me to work out plans to secure it.

I had managed several satisfactory campaigns in Santa Barbara and was well known there. Among my good friends there was George Clyde, who at that time was manager of the local branch of a national brokerage firm. Mr. Clyde was a man of means and of excellent connections. I also knew James D. Lowsley, vice-president and general manager of the First National Bank of Santa Barbara. Therefore it was decided that I should first go to Santa Barbara to see what could be done. Nothing but my compelling belief in Farnsworth could have induced me to undertake a mission to procure money for a scheme so highly speculative as this adventure in television. I had raised millions of dollars for the unfortunate and underprivileged. Now I was seeking a few thousand dollars to prove and support the inventive abilities of one whom I considered a genius.

I first visited James Lowsley at the First National Bank at Santa Barbara. He had been a member of a campaign committee when I organized the community chest of Santa Barbara. I told him the story and let him read Phil's outline of the television scheme. There is something of the gambler in every banker that I have known. Jim Lowsley was no exception. He immediately visualized the possibilities. Of course, it was nothing he could recommend to the bank. But it appeared to him to be a sporting chance for someone who had lots of money.

"It would do no harm to talk to Max Fleischmann about this," he said.

I had met Mr. Fleischmann once or twice but didn't feel that I knew him well enough to tackle him cold.

"I'll call him up," said Jim, and reached for the phone.

Mr. Fleischmann was in and asked me to come right over. On the way over I reviewed all of the fine things Fleischmann had done for Santa Barbara, and remembered how he had been a leader in large subscriptions to the community chest. It all made me a bit diffident in presenting a speculative scheme like our television system to him. I made some remark to this effect as I entered his office. He put me at ease immediately and then listened most attentively as I told my story.

After an hour of interested questioning and discussion, when I felt that I was getting somewhere, Mr. Fleischmann paused and leaned back in his chair in deep thought. After a few minutes he leaned forward and said, "If it were bacteria you were dealing with instead of electronics I would be interested. I wouldn't know an electron if I met it on the street. The whole field, though fascinating and most tempting, is completely out of my range of experience. We have made a success of yeast. Bacteria I know, but electronics—I would be lost in trying to follow what is going on. I'm sorely tempted, but my judgment says that I'd better stick to bacteria."

Before I left he made several suggestions as to who might be interested, but nothing came of them.

When I discussed the matter with George Clyde, he suggested a prominent financier in southern California, a member of a group of capitalists who had set aside a special fund for the furthering of scientific research development. This seemed to be made to order for the Farnsworth scheme, and both Clyde and I had high hopes of getting some support from that quarter.

Mr. Clyde made arrangements for me to meet the gentleman in Los Angeles, and I submitted the memorandum which Farnsworth had prepared. The capitalist stated that he would have it checked by properly qualified engineers and scientists and would give his report in a week's time. Phil was elated at the prospect of getting backing from such a group. It was just the type of foundation which he himself hoped one day to establish.

When I called to get the results of the investigation it was stated that it would be futile to pursue the Farnsworth scheme further because Western Electric, a wholly owned subsidiary of the American Telegraph & Telephone Company, had extensive patents on television and probably had complete control of the new art. This erroneous report is interesting in view of the fact that years later, in July 1937, after a searching analysis of the Farnsworth inventions, the American Telegraph & Telephone Company entered into a cross-licensing arrangement with the Farnsworth company. It is possible that the investigator who made the report had seen the Nicholson patent on a television system which was controlled by A.T. & T. (This patent, as was later found out by Farnsworth, had some broad claims that read generally on almost any electronic television system, but they were by no means controlling in all phases of the art.) This was a most discouraging report, because the Pasadena group of capitalists had access to a fund of engineering advice. In addition, A.T. & T. loomed in my imagination as a powerful company to oppose.

I immediately went out to see Farnsworth and held a council of strategy. For some reason or other, neither of us was greatly disturbed by this report, because it seemed so completely at variance with the Bell Laboratories research program on television which, as far as could be gleaned from the Bell Laboratories journal and other publications, was devoted exclusively to the scanning-disk method. It seemed reasonable to believe that if Western Electric controlled all of the patents on television, they were probably patents covering the mechanical method. We felt strongly that they could not possibly be in conflict with the revolutionary electronic method that Farnsworth proposed.

While not frightened by the report that the telephone company had control of all television patents, the financial outlook was anything but encouraging, because here was an outfit that seemed made to order for the development of the Farnsworth invention, but whose door was closed to us.

Phil felt very strongly about the novelty of his invention, but that did not change the minds of those who controlled the most likely prospect for help that I had yet uncovered. I was frankly worried and told Phil so.

"I've raised lots of money, Phil," I said, "but this is a tougher one than I thought it would be. It's a lot easier to raise money for social welfare than it is to find backing for a speculation."

Whatever my misgivings, there was nothing to do but to continue in my efforts to find money.

"Jim Lowsley is a hardheaded banker," I told Phil. "He is sincerely interested, so there must be a way out somehow. I think I'll go back to Santa Barbara and talk things out with him again."

Upon returning to Santa Barbara, Mr. Lowsley introduced me to a Montecito resident, the son-in-law of the late great railroad builder, James J. Hill. This gentleman, who lived on a beautifully landscaped estate, was an enthusiastic amateur photographer and had had some unusually lovely color pictures printed in the *National Geographic Magazine.*

After some discussion he took me into a large drawing room with a screen at one end. The room was furnished with rows of beautiful French design chairs done in petit point. Here he entertained me for an hour with fantastically beautiful color photographs, probably among the finest that had been produced up to that time.

After he finished showing the slides we went out into the patio by a lily pool. "This is where the nudes were posed," he said. "Color photography is a passion with me. It is yet in its infancy." Then after a pause, "If you can produce television in color I am interested. But if it's just black-and-white reproduction I can't help you."

"I have talked of color with Farnsworth," I replied. "He seems to think color is possible, but as he has described it to me I think it is a long way off. To promise that we could transmit color would be stretching things too far. Our immediate concern is with pictures in black and white."

In this, as in the interview with Max Fleischmann, I was learning much about rich men's interests and hobbies and the limits of such interests. I could not honestly say that we proposed television in color, so this prospect was lost.

Naturally a great many of my friends had taken an interest in my venture in television. All were fascinated by the possibility of the new art, but they were afraid that I had been led astray by a fantastic scheme that had hardly one chance in a million of success. However, there was something about Farnsworth's sincerity and the glamour of the idea of television that made them, against their wills, join in the bizarre hopes that the thing might be a success. In fact, there was so much interest expressed that a business associate and I drew up plans for a syndicate in which a group of our friends were to provide the necessary $25,000. This sum was subscribed on a contingent basis. While both Farnsworth and I felt that such a method of financing was to be used only as a last resort, it did give me some sense of security as I made plans to explore other possibilities. Somehow I felt that $25,000 with no more money in the background was not the kind of backing that was necessary. It seemed to me that it would be much more advantageous to get the backing of more substantial interests. With this end in mind, it was decided that I should go to see some of my friends in San Francisco.

My first thought was to discuss the matter with Mr. Jesse B. McCargar, vice-president of the Crocker First National Bank. Mr. McCargar had been chairman of the fund-raising campaign for Californians, Inc., which I had managed. While I had no thought of being able to get any backing from the bank, I felt that Mr. McCargar's advice on the venture would be helpful.

On the morning of my arrival in San Francisco in mid-August, I went into the Crocker Bank and approached what I called "the throne," on which all the executives of the bank sat.

Mr. McCargar was not at his accustomed place, and one of the attendants told me he was away on vacation and would not be back for a couple of weeks. Since I was due in El Paso on a campaign contract within that time, I was much disturbed and inquired further regarding the possibility of McCargar's earlier return.

James J. Fagan, executive vice-president of the bank, observed my anxiety as I passed his desk and asked, "What can I do for you, young fellow?"

I replied, "Mr. Fagan, I was looking for Mr. McCargar. I had a matter to discuss with him."

"Can't I help you?"

"No, I don't think it is anything that would interest you in the least. It is not an investment, it is not a speculation; it is wildcatting and very wildcatting at that."

This seemed to intrigue Mr. Fagan, who at that time was regarded as the soundest and most conservative banker on the Pacific Coast. In fact, he was one of the very few remaining bankers of the old school. He had lived through much of the California bonanza era. With other bankers he had seen the city through the disastrous days of the earthquake and fire and had participated in its magnificent recovery. He was beloved and revered by the clients of the bank. It was said that he could smell a bad investment before it came in the door, and that he had a sixth sense to detect a souring financial picture, or business venture, long before there was any surface indication of trouble and always had the accounts of such concerns out of the bank months before a financial crisis appeared. On the other hand, where there was personal integrity, industry, and a high sense of obligation, no one was more patient in helping a debtor to work out his financial salvation. It was this latter quality in Mr. Fagan that not only made him a great banker, but one much loved. He was the butt of many jokes regarding the coldhearted, glassy-eyed guardian of the money bags. Everyone said that anything approved by Mr. Fagan had to be good.

"Well, sit down and tell me about your scheme," Mr. Fagan said cordially and with a glint of humor in his eyes. So I sat at his desk and told him the fantastic story of Farnsworth's development as honestly as I knew how.

Mr. Fagan had a habit of drumming the opposing fingers of his hands together as he looked over the rims of his spectacles. With this characteristic gesture he looked at me and said, "Well, that is a damn fool idea, but somebody ought to put money into it, someone who can afford to lose it." I said that I agreed with him and had come into the bank to talk the matter over with Mr. McCargar to find out if there was a possibility of getting someone interested.

Mr. Fagan, in a pensive mood, speaking as his train of thought led him, surveyed the possibility of such backing. "Well, there is Mr. ——, who has more money than anyone has a right to have. He might be interested." Then after a moment's thought he continued, "But he isn't the man for it. He wouldn't have the patience that a thing of this kind requires. There is also a bunch of young fellows down the Peninsula—friends of my son Paul. Those men throw away more money in a year's time than you would need, but they would be unstable and difficult to deal with, so I don't think that would work out."

Mr. Fagan sat in quiet musing for a few minutes. Then he said, "I know a man who sometimes has money to put into ventures of this kind. If you will come in tomorrow about this time I will try to arrange an appointment for you."

The next morning Mr. Fagan sent me over to see an engineer at the branch offices of a large eastern industrial concern. I told the engineer what I had in mind and showed him a copy of the memorandum covering Farnsworth's scheme. Upon introduction he told me that it was not he that had money to invest. The executives of the Crocker Bank often had speculative ventures put up to them, and when such ideas seemed to have merit they occasionally sent them over to him to investigate and give an opinion. In this case he said that Mr. Fagan had asked him to look into the Farnsworth television for him.

The engineer was greatly interested in the Farnsworth memorandum and asked to see young Farnsworth as soon as possible. I arranged immediately for Phil to drive my car up from Los Angeles. In the meantime Mr. Fagan had talked the matter over with W. W. Crocker, son of W. H. Crocker, head of the Crocker family of San Francisco. Mr. Crocker was very much interested and suggested that he would like to have Roy Bishop, an engineer and capitalist, look into the matter. Arrangements were made for a luncheon meeting with Mr. Bishop a few days later.

In the meantime Farnsworth had come to San Francisco. Because of his consuming interest in his invention and his meager finances, Farnsworth, sartorially, was not a very well-turned-out individual. His clothes were shabby and ill-fitting, and generally speaking he had the appearance of a poor inventor. Both Farnsworth and I thought this situation should be corrected, so I took him to a good shop, where he was completely outfitted with a new suit, hat, and haberdashery.

The meeting with Mr. Bishop was to be held at the Palace Hotel, and since Farnsworth had never been at lunch in a hotel of this kind before, he was a bit worried. To allay his feeling of stage fright we took breakfast and dinner at the Fairmont Hotel, which was on the opposite corner of the street from where I lived.

The day of the luncheon arrived, with Mr. Bishop, Farnsworth, the engineer, and me present. Bishop was greatly interested in the television plan as it was unfolded by Farnsworth. I need not have been worried about Phil's self-consciousness, for it completely disappeared as he became engrossed in the discussion of his invention. Mr. Bishop, who had just lost his son, told Farnsworth that one of the things he had done for the boy during his illness was to fit up a workshop where he could experiment on his own ideas. Bishop said that this fact made Farnsworth's ideas doubly appealing to him.

The luncheon discussion continued for two hours and then adjourned to Mr. Bishop's office, where it continued until after five o'clock in the afternoon. Mr. Bishop's experience led him to believe that Farnsworth's estimate of the time and money involved to bring the development to a conclusion was inadequate. He went on to say, "I am convinced that the idea is sound, but doubt your ability to work it out commercially."

From then on it became apparent that in spite of his interest Bishop was reluctant to do anything with the Farnsworth scheme.

Farnsworth has always had some of the fundamentals of a good salesman. As this attitude on Mr. Bishop's part became evident, Farnsworth rose from his seat, walked over to the desk, picked up his brief case and with a courteous gesture thanked Mr. Bishop for his kindness in spending so much time discussing the matter and expressed regret that he could not see the possibilities that we saw in the invention.

As Farnsworth and I were about to close the door behind us Bishop said, "Wait a minute."

He and the engineer held a whispered conversation, after which Bishop said he would like to have an engineer at the Crocker Research Laboratories, by the name of Harlan Honn, look into the matter.

"Honn is a hard-boiled, competent engineer," Bishop explained. "If you can convince him that your proposition is sound and can be worked out, I think we will find ways of backing you."

Honn was called on the telephone and arrangements made for him to meet Farnsworth at my apartment.

Within half an hour, Honn, Farnsworth, and I were in earnest discussion of the television plans. Honn promptly grasped the significance of Phil's ideas.

After he had read the specifications for the system and had had some pertinent questions answered to his satisfaction, he turned to me and said, "Why, sure this system will work. I think very well of it."

There was further discussion, and after dinner I asked Honn to call Mr. Bishop and give him his findings. Later Honn made a formal report in writing to the McCargar-Fagan-Crocker group.

The matter was then held in abeyance until Mr. McCargar's return from his vacation. Immediately thereafter it was agreed by the banking group that if a satisfactory agreement could be reached with the Farnsworth partners, they would take a flier at the Farnsworth scheme to the extent of $25,000. With this arranged among themselves, McCargar asked Farnsworth and me to meet with the financing group in the directors' room at the Crocker First National Bank the following afternoon at two o'clock.

When Farnsworth and I arrived at the bank the group was gathering in the directors' room to talk things over. In the meantime we sat on a marble bench at the foot of "the throne" waiting to be called. Within a few minutes Mr. McCargar came over, put his arm across my shoulder, and said, "I think we are going to back you boys," and invited Farnsworth and me to join the group.

Mr. Bishop acted as chairman of the meeting. He asked Farnsworth to outline the method of procedure he thought necessary to pursue the work he had in mind. In simple terms Phil gave a general sketch of his television scheme, emphasizing in particular the absence of moving parts in the system. Because he was completely self-assured and modest in his presentation, Farnsworth made an excellent impression.

There has always been some quality about Phil that has enabled him to fire the imaginations of those with whom he talks about his inventions. As he talked to the banking group his expressions had a clarity which, coupled with a quality of obvious genius, stimulated the enthusiasm of his listeners without any apparent effort on his part. It became clear that the group would back the invention.

Phil began by saying that he proposed to do for vision what radio was doing so successfully for sound.

"It isn't a magic carpet where whole scenes are sent as a unit instantaneously from one place to another. It is much more complicated than that. We must tear the pictures down into thousands of bits for transmission and then put the pieces together again in perfect sequence at the viewing end. All this must be done in split seconds in order to fool the eye."

"Hasn't this ever been done before?" McCargar asked. "Television is not a new art, is it?"

"No," said Phil, "the concept of television goes back to the discovery of the selenium cell, when they found that the electrical resistance of this metal varied when varying light intensities were focused on it. With this discovery scientists first conceived the idea of transmitting the varying light values of a picture by electrical means. They thought that by scanning an image by some mechanical means they might be able to pick it up a point at a time, transmit the varying intensities one by one, and restore them by a reverse process at the receiving end. No, television is not a new art. Its conception dates back more than fifty years."

"If that is the case," somebody asked, "why hasn't someone made it practical long before this?"

"That is a long story," Phil went on. "Many tried to do something with it. They all attempted to break down the image for transmission by using mechanical devices. The first really halfway practical approach was in 1884, when a Russian named Nipkow, working in Berlin, took out a patent on the 'scanning disk.' In his apparatus he used a rapidly revolving disk with minute holes along the outer edge to accomplish the scanning. In 1889 an inventor by the name of Weiller used a wheel with convex mirrors of highly polished metal on the periphery for the purpose.

"Since the turn of the century the Nipkow and Weiller devices have formed the basis of practically all television experiments. This is what Dr. Ives of the Bell Laboratories is working on. Jenkins of England has succeeded in transmitting a fairly recognizable, crude image with the scanning disk. All of them are trying the impossible. It can't be done by mechanical means. I propose to do it by wholly electrical means by manipulating the speed of electrons."

This was pretty strong medicine for the bankers, but they were used to exploring new fields. They pressed Phil for more details. A newspaper was lying on the directors' table before Mr. Fagan, who sat next to the inventor. Phil reached over and spread it out before him. "If you had a reading glass you would see that this picture of a girl is made up of many small dots. They are produced by a screen on the cut from which the picture is taken. The fineness of the screen determines the quality of the picture. In a good magazine picture the mesh is much finer. There are probably 250,000 dots or units in such a reproduction. To transmit a picture of like quality over television, each of these dots must be picked up separately and sent in sequence. To fool the eye, all this must be done in a fraction of a second. To get smooth motion as in motion pictures, we must probably send the pictures at the rate of 30 a second. In other words, to do the thing successfully we must register and transmit 250,000 variations every thirtieth of a second. That means something like 7,500,000 changes in intensity every second. Such speed cannot be achieved by mechanical means."

"What are you going to use to break down and restore the image?" Mr. Bishop asked.

"I shall use a photoelectric cell to change the picture image into an electron image. Then I will scan the electron image a line at a time with great speed. The picture signal or current will flow out from the scanning anode carrying the electrical counterpart of the light and shade values of the picture to a broadcast unit for modulation onto a broadcasting frequency for transmission through the air. At the receiving end I propose to use a cathode-ray tube, where the scanning process in reverse will be synchronized with the scanning at the transmitter."

"It all sounds pretty complicated and way over my head technically," Mr. Fagan commented to Phil dryly. "I guess we will have to take Engineer Honn's word for it that your scheme will work." Then turning to the others, he continued, "The boy sounds as though he knows what he's talking about, though I can't follow him."

Then turning to the more practical aspects of the moment, Mr. Bishop asked, "How much do you think it will take to produce a recognizable television picture?"

"I believe with $1,000 a month for twelve months I can get a satisfactory result," Phil replied. "But to be on the safe side

I would like to have $25,000 to work with. We may run into some difficulties that I haven't foreseen yet, so we might have to have a little leeway to go on."

"I think it isn't enough," said Bishop, "but I'll take your word for it."

Then there was discussion as to where the laboratory should be established. It developed that a loft at 202 Green Street was used by Mr. W. H. Crocker in some ventures he was backing. These developments were coming to an end, so the space would be available.

"I think the rental there is $75 a month. There's plenty of power handy, too," said Bishop.

Phil enumerated a few simple instruments he felt he needed, and he was told he should get what he needed so long as he stayed within his $1,000 monthly budget.

"Shouldn't we supply this young man with a competent consulting engineer?" Mr. Willis, one of the bank executives, asked.

"I don't think so," Mr. McCargar replied. "This is Farnsworth's show. We are betting on his ability. What he is proposing is not orthodox. What he will do would probably give an orthodox engineer heart failure. This is not engineering we are backing, it is invention. If Phil here succeeds, he must do the thing his way, no matter how cockeyed it would seem to a well-trained engineer. If we back this boy we've got to go the whole way. He must be the boss of what he is doing."

Phil took great encouragement from this, and as the discussion progressed he was led into explaining in considerable detail how he planned to proceed. When he had finished, Bishop turned to Phil and me and said, "Well, young man, and you, Mr. Everson, this is the first time anyone has ever come into this room and got anything out of us without laying something on the table for it." Then turning to the other members of the group he said, "We are backing nothing here except the ideas in this boy's head. Believe me, we are going to treat him like a race horse."

It was agreed that the financial group was to have 60 per cent in the newly formed syndicate in return for acting as trustees of the venture and supplying the funds. The remaining 40 per cent was to be divided between Farnsworth, Gorrell, and me. It was arranged that Farnsworth would be given an allowance of $1,000 a month for the laboratory, out of which he was to receive a salary of $200. As outlined in the earlier discussion, the work was to be carried on in the Crocker Research Laboratories at 202 Green Street, San Francisco.

After the meeting broke up, Farnsworth, Bishop, and I adjourned to the office of Herman Phleger, an attorney, to draw up the necessary papers.

In the discussion it came out that Farnsworth was not of age. Since his mother lived in Utah, it would be necessary for him to have a legal guardian in California, whereupon steps were taken for my appointment as his guardian until he became of age. The formal contract, therefore, was not signed until some time later, owing to the delay in getting the guardianship papers completed.

Phil and I were elated with the setup. It seemed almost too good to be true. Later, as I have told the story to people who knew Mr. Fagan, many have said jokingly that wonderful as was Farnsworth's youthful genius, the fact that Mr. Fagan bought the idea was even more miraculous.

Farnsworth then returned to Los Angeles to arrange for the shipment of his laboratory equipment and to bring his wife and personal belongings to San Francisco. The roadster was pretty well loaded with baggage as Farnsworth and his wife drove north. A large suitcase containing most of their clothing, including Farnsworth's new suit, was on the running board. When they stopped at King City for lunch, someone stole the suitcase. As a consequence, Farnsworth arrived in San Francisco with only the shabby clothes he had put on for making the trip. He hunted up my friend, Harry Cartlidge, and borrowed a hundred dollars to re-outfit himself.

In the meantime I had proceeded to El Paso to fulfill my contract with the community chest in that city. Phil was left to his own devices in San Francisco to get things started. A great weight was lifted from my shoulders by having his idea backed by a strong financial group.

All his life Phil has been a reluctant correspondent. During my six weeks' stay in El Paso I learned how deficient he was in this respect. After my procuring for him his heart's desire in the way of financial backing, it never occurred to him that I might like to have occasional reports on how he was getting on. So the first news I had from him came from Cartlidge, who wrote me that he had loaned Phil a hundred dollars to replenish his wardrobe after the theft. For further details I had to wait until I returned to San Francisco for a personal visit.

By the first of October everything was in readiness to start work at the Green Street laboratories.

The financial backers wisely followed Mr. McCargar's counsel that Farnsworth must be in complete control of his research without the handicap or help of a supervising engineer. They recognized that if his work was to be valuable, it must be the product of genius, rather than engineering, and that he should therefore be in complete command of the enterprise. They had faith in Farnsworth's integrity and industry, and in his enthusiasm for his work.

(((4)))

The Green Street Laboratory

My campaign contract in El Paso made it impossible for me to stay on in San Francisco to be with Phil when he took possession of the quarters at the Crocker Laboratories on Green Street. The word "laboratories" was a flattering name for a large second-story loft over a garage.

The building was directly at the eastern foot of romantic Telegraph Hill. The rear windows of the loft faced the barren rocks of the precipitous side of the hill. High above were apartment houses and single dwellings overlooking San Francisco Bay. It was not uncommon, after severe rainstorms, for loosened rocks to come tumbling down the hillside. On such occasions fragments of stone would sometimes hurtle through the windows or land on the roof with a great clatter.

Theoretically the best location for any research laboratory dealing with television and ultrashort waves, the radio carriers for television, would be some high elevation free from surrounding buildings or hills and far removed from possible interference by electric power lines and motor-driven machinery. The Green Street laboratory was just the opposite of this theoretically ideal spot. The building was practically at sea level and at the foot of a steep hill with power lines all about. Across the street in one direction was a large publishing plant with many motor-driven presses. On the opposite corner was a ship repair plant. However, the roof of the lab had a clear view of the Bay to the east and a fairly clear outlook on downtown San Francisco to the south.

The initial equipment and personnel of the Farnsworth laboratories were a few hundred dollars' worth of electrical equipment and experimental apparatus shipped up from Los Angeles, a long bench, a desk and a few chairs, and the inventor, Philo T. Farnsworth. Out of this little nest egg was to be hatched a complete electronic television system.

It should be remembered that at this time Farnsworth had just reached his twentieth birthday and that he had never seen the inside of a research laboratory, or, for that matter, any large electronic manufacturing plant.

The development of the tubes he had in mind was to require the greatest skill and subtlety known to the glass-blowing art. The timing of the pulses in the electrical equipment to be devised must be accurate within less than one seven-millionth of a second. Since the speed of electrons is nearly equal to that of light, Farnsworth felt that they must naturally be amenable to a corresponding astronomical precision of control. In producing a picture it would be necessary to manipulate electrical strength represented by single electrons to produce amplification to the millionth power.

Farnsworth was conscious of the problems facing him, but he did not let them trouble him. He attacked the whole assignment with no engineering experience and little engineering knowledge, but to compensate for these inadequacies he had courage and genius. The courage was not the foolhardy type born of ignorance. His was the real courage of the pioneer who knows the goal but has little knowledge of the intervening terrain. He had no idea that the problem was going to be as difficult and complicated as it proved to be, yet he had the confidence to believe that his conception could be made to function successfully.

Since there were no glass-blowing facilities on the Pacific Coast equipped to make the transmitting tube that Farnsworth dreamed of as the core of his television camera, the first step he took was to send a wire to his brother-in-law, Cliff Gardner, to come to San Francisco immediately to set up the glass-blowing laboratory. Gardner's total training for this was a high-school education, a boldness comparable to Farnsworth's, dogged perseverance, and no knowledge of the subject.

The explanation of this strange choice of an assistant lies in three traits of Phil's character that have persisted throughout his career. First, his Mormon training made him cautious, and in getting Gardner to help him he was sure he had someone whom he could trust implicitly. Second, he was afraid to employ engineers or technicians who were better equipped technically than himself. Third, he always liked to have members of his own family associated with him in his work. All these traits were serious handicaps to Farnsworth in his development work. Often he and an inferior assistant spent months in solving problems that were at the fingertips of more fully trained engineers. This quality was a source of both strength and weakness. It contributed greatly to his originality of thought, but it retarded his ability to get things done in a practical way.

In due time Gardner arrived, and with consulting help from a member of the University of California faculty, work was begun on building a vacuum pump and glass-blowing table for tube construction. As soon as the tools were ready, Gardner got down to the business of learning how to make electronic tubes. He studied with avidity all of the literature on the subject and after months of painstaking effort was able to make a beginning in the art of building the tubes required as the heart of Farnsworth's television system.

When Gardner got into the work, he found that the vacuum pump initially constructed was not adequate for the difficult job of building a television camera tube. Therefore, with Phil's consulting assistance, he set about building one that would meet their needs. It was fearfully and wonderfully made. We joked about its being something copied from one of Rube Goldberg's cartoons. Fantastic as it looked, it did the work.

Electric ovens had to be included to heat the elements in the tube while on the pump to drive off impurities in the metals. This was necessary to prevent troublesome stray gases from developing in the tubes to spoil the vacuum. While this equipment and procedure was commonplace in any well-established electrical research laboratory, it was all new to Phil and Cliff Gardner. They had to learn what tools were necessary and build them as they went along.

During Gardner's learning period Farnsworth was particularly fortunate in enlisting the assistance of Herbert Metcalf, a radio engineer and physicist who had had some experience in cathode-ray tube developments with Dr. Joel Kunz at the University of Illinois. He had also had experience with photoelectric cells built by Dr. Kunz for the use of Metcalf's father and the astronomer Joel Stebbings in measuring the light of stars.

Metcalf and Cummings, the consultant from the University of California, built the first transmitter tube for Farnsworth to use for testing purposes, thus enabling Farnsworth to carry on the other avenues of his experimental work while Gardner was learning the art of glass blowing.

To assist him in building the radio circuits two radio technicians were employed and set to work by Phil in the design of the transmission and receiver apparatus. They were not trained engineers. They, like Gardner, learned of the problems in television as they went along.

The transmitter tube, which was to break down the picture and convert it into an electron image for transmission, was Farnsworth's first concern. As I have already mentioned, he conceived it as a cylinder about the size of an ordinary Mason jar with an optically clear window sealed into one end and a photosensitive plate at the other end. As they got into construction on this tube, two major difficulties were encountered. One was the art of sealing in an optically clear flat disk of glass at one end of the tube without leaving stresses and strains in the glass that would cause cracks under hard vacuum when the tube cooled and was subjected to varying temperatures. The other was the distillation of the photosensitive substance on the plate at the rear of the tube. Incidental to these two major problems was the difficulty in getting the right type of electrical leads in and out of the tube so that it would operate as the inventor had planned.

Phil decided that the tube was to be known as the Farnsworth "dissector" tube. This name was chosen because it was descriptive of the process that took place in breaking down the image for transmission.

Potassium hydride was used as the photoelectric surface in the early dissector tubes. Since potassium combines with oxygen when exposed to air or water, the potassium pellets came from the suppliers submerged in kerosene. Gardner had to learn the art of distilling the pure potassium from the commercial pellets and sealing it in airtight glass tubes for future use. To deposit the potassium on the rear plate of the tube as the foundation for the photoelectric surface, in building the dissector tube the capsules of potassium were sealed to the lead into the tube and driven in by the application of heat. Then hydrogen gas was introduced in such a way as to insure its combination with the potassium surface to form potassium hydride. This gave a good photosensitive surface. It was an intricate and exacting operation, better suited to the abilities and background of experienced chemical engineers than to the capacities of a boy fresh from high school.

The fact that Farnsworth and Gardner devised the technique and successfully built the tube that the inventor had in mind is a tribute to their ingenuity and perseverance, and to the tutoring of Herbert Metcalf.

Such problems delighted Phil. He reveled in the simple laboratory facilities provided for him and, like a boy with a new mechanical toy, enjoyed showing "house of magic" stuff to the uninitiated. Liquid air, for instance, was one of the essentials for operating the vacuum pumps. When visitors came to the lab Farnsworth took boyish pleasure in removing the cap of the liquid air container and inserting a piece of rubber hose for instant freezing. Pulling it out, he would break the brittle frozen rubber. Igniting pellets of crude potassium by throwing them into water was another of his favorite stunts.

Instinctively Phil enjoyed enlisting the interest of people in his ideas. There was something of the evangelist and propagandist in his make-up. He delighted in people's approval and interest in what he was trying to do and used these simple showman tricks to arouse it.

Farnsworth spent a great deal of his time in studying the mathematics involved in his invention. He had a peculiar genius for thinking in mathematical terms as clearly as the ordinary student thinks in terms of language. He also had a gift for visualizing the behavior of electrons within a vacuum tube. Though all of the actions of electrons were completely invisible, he seemed able to see them as one would a swarm of bees buzzing about. His mathematical ability came into full play as the physical setup for the television system began to take tangible form. He would literally devour a new branch of mathematics in one gulp if it was needed in the solution of the problem at hand. He was a tireless worker. Mr. McCargar often said during the early days, "Phil Farnsworth hasn't a lazy bone in his body."

Phil followed through on his self-imposed regimen of schooling himself in constructive thinking. He had great faith in the ability of the subconscious mind to solve difficult problems. When completely stumped in his effort to see his way clear to the solution of a problem, he would purposely postpone it to let his subconscious mind work on it. Phil often said that he charged his mind with a knotty problem just before falling asleep, and then set the alarm to awaken him the following morning an hour before rising time. This extra hour would be spent in bed in quiet thinking. Usually this would bring the solution of the problem that had occupied his attention and baffled him the previous day.

It was at this time that Farnsworth recognized the importance of the design of the focusing and magnetic coils which were to control the electron image within the dissector tube. The old ones that I had wound in such a daubing mess were far from adequate. Fairly good facilities were installed for winding new coils that really looked like something.

His original patent application stated that he would use electrostatic focusing of the electron image within the dissector and would deflect it for scanning by electrostatic plates. However, when he set about to make up the equipment he decided to return to the use of magnetic coils for focusing and scanning.

In high school the study of optics had always fascinated Phil.

The works of the Irish scientist, Hamilton, particularly engaged his attention; therefore, when Phil became interested in electronics, he carried over to it some of the ideas gleaned from the field of optics. He conceived the idea that he could focus his electron image in the dissector by use of a coil acting as a magnetic lens. Similarly he believed that he could control the scanning by use of magnetic coils to attract and repel the electron image. By this orderly attraction and repulsion he would sweep it across the anode of the dissector a line at a time. Two sets of coils would be used, one to move the image rapidly from side to side, the other to move it up slowly so that each line would be brought in place for scanning by the more rapid lateral motion.

In his original conception of the magnetic lens Phil thought that as the current increased the focus would become sharper, but early experiments showed this was not the case. The focus sharpened to a certain strength, but then if more current was applied the image again became blurred.

This magnetic lens was another original contribution to electronics. It provided Farnsworth with an important patent in television scanning.

The design and placement of the magnetic coils around the dissector tube became an essential factor in the production of a clear, undistorted image. In his early experiments Phil was not fully aware of the significance of these factors. Later they remained to plague him with what was called a pincushion effect and an "S" distortion across the field of the pictures. These requirements came later, but in the beginning he was concerned principally with the timing of the pulses and their steadiness to get an orderly scanning of the image.

As he got into actual work two practical problems presented themselves. The first one was that of the sensitivity of the photoelectric surface in the dissector tube. Experiments proved that wide variations existed in the number of electrons released by different metals, or combination of metals, in the photoelectric process. In other words, he found that while a nickel alloy gave off electrons when light was focused on it, the number of electrons per unit of light intensity was much less than the number released by other metals.

Potassium hydride served well enough as the photoelectric surface in the first dissector tubes when Phil and his helpers were striving to prove that the system would actually transmit an image. That simple fact had to be proved before he went on to any refinement. However, he knew from the beginning that to achieve satisfactory results development of more sensitive photoelectric surfaces must be one of his major lines of investigation.

After constructing several dissector tubes with the potassium surface, he determined to change to cesium oxide, but here again he found that there was great variation in the electron emission of seemingly identical cesium oxide coatings. The emitting surface was a most baffling and elusive problem. It was destined to harass him and his assistants for many years before the production of cesium surfaces approached uniformity and dependability. By some strange luck most of the earlier dissector tubes possessed remarkably good sensitivity.

The second problem, and one that was equally baffling, was that of getting an amplifier of sufficient power and stability to step up the infinitely small electrical currents produced by the scanning of the image within the dissector tube.

In 1926, when Farnsworth began his work, radio technicians were just beginning to realize the importance of amplification in getting high-fidelity reproduction in sound. The amplifiers were crude and of low power in comparison with those in use today. In the original conception of his scheme Phil had no means of knowing how hard it would be to overcome the sensitivity problem and to produce adequate amplifiers, nor did he realize what a fertile field of research lay before him in the solution of these difficulties. He did know that if he was to succeed with his plans he would have to find more efficient means of amplification. Even before he was well settled in the San Francisco laboratories his inventive mind was searching out new and unorthodox methods. In working the problems through he came to recognize the importance of their solution in other fields of electronics.

(((5)))

Crude Beginnings—the First Television Picture

SOON AFTER Farnsworth was established in the Green Street laboratories it became obvious that it would be too expensive and quite unsatisfactory to have the patent work in the hands of Los Angeles attorneys. Consequently the patent file was transferred to the offices of Charles S. Evans in San Francisco.

Donald K. Lippincott, a graduate of the University of California in radio engineering, and former chief engineer for the Magnavox Company, handled the radio patent work in the Evans office, so that the Farnsworth account naturally fell to his lot.

Shortly before that Mr. Lippincott had been consulted by his friend and engineering associate, Herbert Metcalf, regarding his reaction to Farnsworth's ideas. Lippincott looked upon them with cool indifference as being the fantastic dream of a visionary youth, his main objection being that there were no usable radio channels broad enough to carry the television signal required for adequate detail in the received image. (The ultrashort waves now assigned to television were then entirely out of the range of practical engineering as channels for radio transmission.)

When the Farnsworth account was turned over to Lippincott in the Evans office, he naturally made a close study of the inventions. It was not long before he became an enthusiastic convert to the Farnsworth ideas of television, and there grew up between Farnsworth and Lippincott a fast friendship and mutual respect that contributed greatly to the success of the Farnsworth venture. Lippincott had an active interest in science for its own sake. This, combined with an amazingly retentive memory and a fine sense of humor, helped to endear him to Phil as a friend and confidant.

Lippincott, a man of unusually broad learning and mathematical ability, has often remarked that early in their association he found it necessary to give Phil help in his endeavors to master the branches of mathematics required in solving certain television problems. He relates that in a surprisingly short time Phil had gone far beyond him in his knowledge of mathematics and that he found it difficult to follow him.

Not long after the Farnsworth account was turned over to the Evans office, Lippincott established patent offices of his own. Naturally Farnsworth wished Lippincott to continue handling his work, as did Mr. McCargar and the other backers of the venture. As a result, the account was transferred to the office of the new firm, Lippincott & Metcalf, and Farnsworth continued to have the advantage of Lippincott's friendship and counsel as his patent attorney. Since Herbert Metcalf was the other partner in the firm, Phil also continued to have the advice and consulting help of this very practical scientist.

Lippincott's knowledge of electronics enabled him to see the value of Farnsworth's research work not only in relation to television, but in its application to other devices of radio. Consequently, Farnsworth's research into the problem of amplification and of improving the sensitivity of the dissector tube became very fertile fields from which Lippincott gleaned much valuable material for patent applications. This resulted in extending the Farnsworth patent structure from its original idea of one broad patent covering his television scheme to a whole mass of interrelated patents reaching into many fields of electronics.

As Phil progressed in the solution of his problems the Lippincott office became increasingly busy with the filing and pursuit of patent applications in the Patent Office for Farnsworth and his associates. The Farnsworth laboratory came to be recognized as an important electrical research laboratory. Phil had an unusually good nose for patentable material. With the assistance of Lippincott's broader experience he developed a practical sense for what was novel in the electronic field. From the very first, therefore, he laid the foundations for a broad and sound patent structure.

After the first dissector tube was completed and the magnetic and scanning coils adapted to it, there came the problem of the design and building of the necessary electrical equipment to generate the wave pulses for scanning the image. It had to be built from scratch by the cut and try method. Neither Phil nor his technicians had a broad orthodox training in electrical engineering, so they were not handicapped by orthodox procedure.

One set of coils was necessary to control the oscillation of the electron image back and forth laterally so that it could be picked up a line at a time by the anode in the front of the dissector tube. The other pulse was necessary to swing the picture up and down at a stated frequency of so many times per second in order that the lines could be picked off in regular sequence and a sufficient number of pictures could be scanned to give the illusion of motion.

In developing the circuits required for this accurate scanning work, as well as for the construction and testing of tubes, a variety of meters and test equipment was necessary. Since neither Phil nor Cliff Gardner had had any technical experience up to this time, it was necessary for them to find out not only what equipment was needed but also to learn how to use it effectively. This was no mean assignment. They had to feel their way and learn as they went along. It was here Phil first found that while $1,000 a month for laboratory expenses looked large before he was located in the Green Street quarters, when it came to stretching it over salaries, rent, radio parts, batteries, chemicals, and so forth, there was not much available for high-priced test equipment.

The result was that from the very first Phil never seemed to have all of the facilities the task required. He often spoke of this as an advantage, since it made for resourcefulness and invention and often led to a simplicity and directness of approach to a problem that might otherwise have become too deep.

Once the transmitter circuits and tubes were under way, Phil attacked the problem of the receiver set. The picture was to be received on the fluorescent surface of a specially designed cathode-ray tube. The ordinary cathode tube in use at that time could not meet the requirements for the television receiver, so Farnsworth made a design for a pyrex glass envelope, which he sent to the Corning Glass Company in New York State. In due time a limited supply of these glass envelopes was obtained as the basis for building the tubes upon which the image was to be received.

With these pyrex blanks as a foundation, Gardner was put to work to develop a tube for the receiving end of the system.

It followed the exaggerated pear-shape design common to all cathode-ray tubes. The stem of the pear was a narrow cylinder; the other end was a bulbous flask with a nearly flat surface for the reception of the picture. It was necessary to coat the inner side of the flat surface with a fluorescent material that would meet the requirements for the reproduction of a television picture. This meant that it would have to be very sensitive to the bombardment of a beam of electrons and that each point of the surface must be able to turn the light on and off in a split fraction of a second to give clarity to the image. In other words, once the point had been bombarded by electrons, it could not hold the glow over any appreciable period without destroying the detail of the reception.

Many silicate combinations were used for experimentation. Finally willemite was found to be the most effective. Great pains had been taken in making the compound, and batch after batch was made before satisfactory results were secured. In one instance a special grinding apparatus was set up in order to reduce the willemite particles to the needed fineness.

The design of the electron gun within the stem of the tube was an equally difficult and delicate problem. Before Farnsworth started his research work no cathode-ray tube had been developed that focused an electron beam with sufficient sharpness to etch a well-defined fluorescent image on the willemite surface. Here the Farnsworth genius, and a great deal of patience on Gardner's part, produced the electron gun and focusing coil that gave a satisfactory tube for experimental use.

Finally a usable receiver tube was developed. This was christened an "oscillite" by Farnsworth because it produced a glowing image on the end of the cathode-ray tube by the oscillation of a beam of electrons under the influence of the focusing and scanning coils.

Once the receiver tube was completed, it was necessary to build the remainder of the receiving set. This required apparatus to synchronize the scanning coils at the receiver with those at the transmitter, and amplifiers to build up the received signal to the required strength for introduction into the cathode-ray tube for scanning.

Naturally at this stage all transmissions were by wire. No effort was made to send a picture by radio. Also in the early stages of experiments the synchronizing pulses were sent over a separate line. Developments necessarily proceeded a step at a time. The first effort was geared to prove that an image could be transmitted by wholly electronic means. Wired transmission was used because it was the simplest.

After all these problems had been worked out painstakingly one at a time, they were to be put together in the first edition of the Farnsworth television system. In one room the dissector tube with its coils and amplifiers was placed on a small stand before a windowlike aperture in the room. The room itself was copper-lined. The dissector tube was hooked up to some panels containing the scanning generators. Leads were fed into a black box containing the amplifier. A copper tube led out of the amplifier into the receiving room, where another box contained the receiving tube and the necessary receiving set apparatus. It was all very handmade and crude-looking. Then the tests were begun.

It must be remembered that up to this time the best television work that had ever been done in the world had been accomplished by Baird of London with his mechanical scanning-disk apparatus. He had succeeded in transmitting an image of what was termed 40-line fineness, which really meant a very blurred sort of picture, owing to the lack of detail. A 40-line picture would mean that there were only 1,600 elements in the image, as compared to the minimum of 250,000 elements now considered necessary in television transmission.

When Phil connected up his first transmitting and receiving apparatus in the fall of 1927, he had no illusions about the quality of the image it would produce. It was possible that the whole scheme wouldn't work and that no picture would be transmitted. If an image was produced it was expected that it would be nothing but a crude outline of the simplest sort. The first problem was to prove that electronic scanning would work. The transmission of any picture, however simple, would suffice for the test.

Phil chose the most elementary image for the first trial. He painted a black triangle on a clear piece of glass for the initial tryout. I knew what was going on and asked Phil to let me be present. After much adjusting and days of planning, Phil phoned that all was in readiness. I went to the lab. We all felt something of the historic importance of the occasion and were keyed up in anticipation of what might take place.

I went into the room that Phil used as his office. He was doodling over some electrical circuits.

"I think we will have a picture as soon as the boys get the new circuits wired up," he said. "It won't take long." Then after a pause, "I don't know how good it will be. The signal is very low, and we may not be able to get it out over the noise."

After a few minutes we strolled out into the lab. Cliff Gardner was tinkering around the crude boxlike television camera, and the other boys were in the receiving room fussing around with the amplifier. The light source was a carbon arc. Finally, when all seemed to be ready, Phil took a glass slide with a black triangle painted on it and laid it beside the camera.

"This will be our first picture," he said.

Cliff Gardner stayed at the transmitter. Phil and I went into the receiving room. The cathode-ray tube with its auxiliary apparatus was mounted on an oblong box of imitation mahogany. Here was our first television receiving set. We watched as a square luminescent field of bluish cast appeared on the end of the receiving tube. A series of fairly sharp bright lines was unsteadily limned on the screen, which was about four inches square.

"Put the slide in," Phil told Cliff.

Cliff did so. The luminescent field was disturbed and settled down with a messy blur in the center. By no stretch of the imagination could it be recognized as the black triangle that we were supposed to see. Phil and I looked at the blur with a sickening sense of disappointment.

Phil suggested some adjustments on the amplifier and the scanning generator circuits. There was a lot of feverish puttering around with no improvement in the results. I felt that I was making the fellows nervous, so I went back to Phil's office to wait. Phil was so certain he was going to get results that I didn't have the heart to leave the lab.

Finally, after a couple of hours of struggle, Phil came to the door and announced, "I think we've got it now."

We again went into the receiving room. Things were turned on again. The bluish field lighted up. Cliff put the slide in again. A fuzzy, blurry, but wholly recognizable image of the black triangle instantly filled the center of the picture field. This was our first television picture! Phil and I gazed spellbound for a while and then with a deep sense of satisfaction shook hands silently.

This interlude of satisfaction endured but for an instant. Then Phil burst forth with a shower of ideas, telling the boys in hurried, feverish words of changes to be made.

I was greatly encouraged. Poor as the results were, we now knew that the principle was sound. It was visual proof that we were on the right track.

It had been my chore to keep Mr. McCargar, Mr. Fagan, and the other backers of the venture informed regarding the progress being made at the laboratories, since all thought it best not to visit Phil until he had some definite evidence of progress to show them. Whenever I returned from an out-of-town campaign engagement I would first visit the laboratory and then go into the Crocker Bank to have a chat with Mr. Fagan and Mr. McCargar regarding Farnsworth's progress.

Very often I would be greeted by Mr. Fagan, who would make rings around his eyes with his forefingers and thumbs. Peering through them he would ask, "Have you seen any dollar signs in that Farnsworth tube yet?" As a result I thought it would be appropriate to transmit a dollar sign so I could tell Mr. Fagan I had actually seen one in the tube.

When the excitement had died down a bit after the transmission of the triangle and the boys had made some of the adjustments that Phil had suggested, I told Phil and the rest about Mr. Fagan's dollar sign.

"Can't we paint a dollar sign on a piece of glass and use it instead of the triangle on the next try?" I asked.

"Sure," said Phil. "Cliff, can you fix one up?"

In a few minutes the dollar-sign slide was ready. Again we went in to see what would happen. As the slide was put in, the dollar sign fairly jumped out at us on the screen. The changes that Phil had suggested really did make a difference. The image was much less fuzzy and more clearly defined.

The next morning I dropped in at the bank and told Mr. Fagan and Mr. McCargar that I had actually "seen a dollar sign in that tube." I didn't gild the lily, so both of the bankers decided to wait until Phil was a little farther along before they went down to see his picture.

When I visited the laboratory a few days later I found that Phil was using some negatives of photographs of Pem, his wife, and of Cliff Gardner. They had little density so far as light values were concerned. As a consequence, they televised much better than the solid black triangle and dollar sign.

"Here is some real progress!" I thought. I was so fascinated that I stayed around all afternoon.

Later Cliff was puttering around the camera with the field left on. I was at the receiver waiting for what might show up next when I saw a wraithlike cloud pass across the picture screen. As I stood there it would fade out and then come back on the field. I was puzzled as to what it was. Then for a second the end of a smoking cigarette came into the lower right-hand corner of the field. It was very distinct. I knew now what the cloud effects had been. They were the smoke from Cliff's cigarette as he worked bent down in front of the television camera. I called to Phil to come and look, but Cliff had moved away.

"Cliff, smoke your cigarette in front of the camera," I yelled.

He obliged, but the heat from the arc light made it difficult. First he tried the end of the cigarette, and smoke curling from it, and then the profile of the mouth and nose of the smoker were brought into the picture.

One day when I was at the laboratory they tried the transmission of a pair of pliers in my hands, but the picture of the pliers did not work out so successfully as the smoker, owing primarily to a lack of experience in lighting and handling sharp contrasts at different focal strengths.

The fact that any image at all could be transmitted by this

crude apparatus was positive proof to Farnsworth and me that his television system would work. Our enthusiasm grew with each new improvement in the image.

The most convincing picture at this time continued to be the one of the cigarette smoke. The smoker had to stick his nose close to the transmitter tube, and because the lights were very bright, care had to be taken to avoid a blistered nose. However, this small success served to bring home emphatically the magnitude of the problems ahead. The question of sensitivity and of amplification became major considerations, and plans were made to develop the necessary apparatus.

The amplifier developments were the most heartbreaking ones. I would leave San Francisco with high hopes of Farnsworth's latest plans for an amplifier that would be satisfactory, only to return to learn that much had been expended in work, salaries, and materials with results that were desperately disappointing.

The black box in which the amplifier was housed came to be a Jonah—or, to change the figure, a yawning maw into which thousands of dollars were thrown without results. Upon one occasion when I returned from an out-of-town trip, I found the laboratory staff ripping out the contents of the black box and salvaging such tubes and equipment as could be used again. This was all that was left of months of work and $5,000 in funds. There seemed no end to the building and discarding of these expensive experimental amplifiers, but finally, after months of tedious work and the expenditure of thousands of dollars, one was operating "flat to 300 kc." as the engineers say.

Previous to the advent of television there was no need for the broad-band amplification that is the first requisite in picture broadcasting. Sound broadcasting was successful without it, and consequently radio engineers had not concerned themselves with the problem. Therefore, no amplifiers were yet developed that would meet the demands required by Phil in his experiments. The "black box" was the first step in the long road to adequate amplifiers for transmission of the broad band of frequencies necessary to produce the full detail in a television picture.

It was this problem of amplification and broad wave band requirements that led Lippincott in his first analysis of Phil's plans to the conclusion that the whole scheme was fantastic and could not be made practical. I think in the beginning Phil was not wholly aware of the difficulties in this regard. With my limited knowledge I was blissfully ignorant of this seemingly hopeless limiting factor. Had I known as much about it as I do now, I doubt whether I could have gone into the scheme as wholeheartedly as I did.

As the problem loomed in Phil's mind, he attacked it with great courage and initiative. Lippincott followed along, confident that between Phil's genius and the rapid developments in electronics a solution would be found. On the part of both of them it was a splendid example of faith and courage.

In his efforts to develop better and better amplifiers, Phil, as usual, attacked the problem on several different fronts. He felt that if he could devise some new means of controlling the action of electrons within vacuum tubes he might find what he needed. Therefore, he studied the action of electrons. Among the phenomena that came to his attention was that of secondary electron emission, in which new electrons were released from the surface of metals when bombarded by primary electrons. Out of this grew his multipactor tubes, which came into prominence several years later.

At this time he began to work on what he called an "admittance neutralized amplifier," a device which became very useful. Both of these principles were revolutionary when he first started experimenting with them. In the development of amplifiers, necessity truly became the mother of invention for Phil. Out of it grew several practical and valuable patents applicable to the broad field of radio.

(((6)))

Televising Motion Pictures

THE LACK of sensitivity in the transmitter tube made the pickup of objects from life very difficult. Farnsworth therefore determined that the next step in his program was to transmit motion-picture film successfully. This presented many and varied problems. The characteristics of the scanning process of a television transmitter tube and the operation of a motion-picture projector had to be reconciled. It was found that it was not feasible to transmit motion pictures by the use of the "shutter" type of motion-picture projector.

To those unfamiliar with the operation of the motion-picture projector it is necessary to explain that the action is produced by flashing onto the screen, at the rate of twenty-four pictures per second, the individual frames which make up the film. The film is fed into the projector a frame at a time at the rate indicated. Each separate frame stops momentarily while it is flashed on the screen. While the next picture is being brought into place the screen is darkened by a shutter, so that during two-fifths of the time in the ordinary projection of cinema film the screen is dark. The eye, which continues to register the image for a fraction of a second after it disappears, does not perceive this interval of darkness.

When the shutter was down and the screen darkened, the scanning went on over the darkened surface, a line at a time in the same manner as over the image. This brought intermittent black flashes on the television receiver tube which gave to the eye the appearance of a violent flicker. The fact that the pictures were scanned at the rate of 30 per second while the projector ran them at the rate of 24 per second added confusion and made the flicker more unbearable. Consequently, to televise them successfully means had to be devised to scan the picture with the film running continuously. Since the picture was transmitted one scanning line at a time, the shutter could be eliminated. With this done, the film flowed evenly across the scanning area and the picture elements were transmitted for reproduction to the received without interruption. This eliminated the major difficulties of the flicker. The operation required nice timing and precision. Here again Farnsworth's great ability in thinking and actually visualizing an operation in mathematical terms made it possible for him to develop a new system to satisfy his requirements.

The purchase of an old motion-picture projector for use in the experimental setup was one of the first times when it became imperative for Phil to overstep the limit of $1,000 set as his monthly allowance. The staff of the laboratory also had to be increased to meet the expanding program of research. In this, as in subsequent enlarged expenditures, the financing group were lenient and co-operative.

While work was going on in the reconstruction of the motion-picture projector, changing and testing was carried on day in and day out throughout the whole embryo system. The black figures painted on glass, the photograph negatives, and cigarette smoke were discarded for more complete images. Better lighting, as well as amplifiers and dissector tubes, contributed to the improvement. A special wide-angle lens was needed in the television camera to project the image onto the photoelectric surface in the dissector tube. The procuring of the best lens for this purpose, and its proper mounting, also required much patient testing.

I returned to San Francisco from a campaign assignment and visited the laboratory to find the operators about ready to transmit motion pictures. They had been testing for days. I was told that operations would probably begin the following day. Needless to say, I was there the next afternoon to see what they had to show. The motion-picture projector was an old rattletrap secondhand machine. It had been overhauled and the gears retimed to meet Phil's requirements. As for still pictures, a carbon arc was used as the source of light. As I stepped into the copper-shielded transmitter room everything was in readiness. After explaining the setup Phil took me into the darkened receiver room where we viewed for the first time the actual transmission of a motion-picture film by electronic television.

The picture was one of a hockey game. The players were quite blurry, but one could definitely see them as they scooted from one end of the rink to the other, or as they sprawled in the rough-and-tumble before the goal. The rink was brilliantly lighted. The contrast between it and the dark figures was excellent. I was particularly delighted with the clarity and precision of the motion, though in the first transmission of the film it was not possible to see the puck.

We ran the film through several times. Here, I felt, we had made another giant stride in our progress toward practical television.

At first not all film was usable for our television purposes.

Farnsworth secured several reels from a local film exchange and ran them through to pick out excerpts that were "telegenic." Through a North Beach motion-picture house one of the boys got hold of a bootleg film of the long-count episode in the Dempsey-Tunney fight at Philadelphia. It was used principally for laboratory demonstration purposes. Another piece of popular test film was an excerpt from the Mary Pickford–Douglas Fairbanks picture *The Taming of the Shrew,* in which Mary Pickford combed her hair at least a million times for the benefit of science and the development of television.

This loop of film was especially suitable for test purposes because of the detail in the picture. In the beginning Miss Pickford's features and the lovely glowing strands of her hair were brought through reasonably well defined, but the background, such as the items of her attire and the frame of the casement window, were not so readily distinguished. It was fascinating to see the improvement from day to day, from week to week, and from month to month by the amount of detail that could be seen on the screen as this loop was scanned.

(((7)))

The Growth of Speculative Value

SHORTLY AFTER Gorrell's and my return from the El Paso and Tucson campaigns late in 1926, Gorrell married and decided to give up the roving life of a campaign organizer and settle down in San Francisco. Here he secured a job as research analyst for a brokerage house and got into the swing of the dying gasps of the bull market in the era of beautiful nonsense.

One day a year or so later he took one of his former Stanford schoolmates, who was now a stock salesman, down to the laboratory. The stock salesman was immediately set on fire by what he saw. He visualized its great speculative possibilities and set about doing something about it. The first I heard of it was in a letter which Gorrell wrote to me in Santa Ana, where I was managing a campaign, telling me that he had sold a tenth of his 10 per cent of the Farnsworth company for $5,000. This was heady stuff for Gorrell. Some time later, when I returned to San Francisco, I visited him in his office. He informed me that the price of his holdings had doubled since he last wrote me, and that he was going to sell a substantial portion of what he had and try to make a real killing in the market. The market broke badly for him, and he lost heavily. As time went on he disposed of all his Farnsworth holdings at what appeared to him to be attractive prices.

In the meantime Phil had taken advantage of the situation and sold enough of his holdings to purchase a very comfortable house on the Marina facing the Presidio and to furnish it quite handsomely. From this time on Phil supplemented his modest salary by additional funds realized from small sales of his capital holdings. This was Phil's first opportunity to have anything beyond the simple necessities of decent living. When he first started work in the laboratory on Green Street he had found a modest place in Berkeley, where he and Pem stayed for a while. Then he moved to a rather bleak apartment in the Hyde Street area of San Francisco. My most vivid memory of the place was that it had a fireplace that smoked outrageously whenever Phil and Pem tried to use it.

When a little more money was available, Phil arranged for his mother to come and live with him in the new house in the Marina. Later she acquired an establishment of her own. Phil was always thoughtful of his family's welfare, and in due course all of his immediate family were established near him in San Francisco. He was reticent about the extent of financial help he gave them, but it was generally presumed that it was considerable.

Pem particularly enjoyed having a home of which she could be proud and a few of the simple luxuries and niceties in dress that she had longed for. She was a lovely girl and is now a beautiful woman. She had instinctive good taste and fine feelings for gracious living. Phil was inordinately proud of her and was happy to give her the means to dress well and have a fine home.

When word got to Mr. Bishop, who was acting for the financing group in their relations with Farnsworth, Gorrell, and me, that some of the interests had been sold he tallied up the amount the group had spent to date on the television development. Finding that it was more than double the $25,000 originally agreed upon, he called the three of us into his office in the Crocker Building and told us that under the circumstances he felt that we should now bear our pro rata share of the assessments for the developments. In fairness we had to agree with him, and from that time on we put up our portion of the funds required.

More and more news of what we were doing got noised abroad in the San Francisco financial district until practically all of the investment houses had knowledge of what we were doing. As a result it was not difficult to find men who were willing to take a flier in this highly glamorous speculation when we needed money to meet expenses.

(((8)))

Clearing the Image of Smudge and Blur

OBSERVING advancement in the quality of the picture over a period of time could be compared to adjusting a pair of binoculars to view a distant scene. As the adjustments were made the picture was brought more sharply into focus. It was the day-long, day-after-day patient experimenting, adjusting, refining, and testing of the electronic apparatus that brought increasing sharpness to the image.

Farnsworth and his assistants were pioneering in a new field. What they were doing had never been tried before. In their work his helpers were led into strange new fields of electronic exploration. Here the working of Farnsworth's mind was at its best. He never attacked a problem in the orthodox fashion, but always found some new way of startling originality to achieve results.

Farnsworth in these days usually appeared at the laboratory at about ten o'clock in the morning. His entrance was like a fresh wind bringing an argosy of new ideas to be tried out by his assistants. In conference with them and Lippincott, Farnsworth's ideas were often found worthy, but more often fallacies would become evident and the ideas would have to be discarded. But it was the repeated method of trial and error with new and untried ideas over months and years that assured the slow but certain progress in the improvement of the television picture.

Film for test purposes had a definite life span. The Mary Pickford film was the favorite, and as a result scores of prints from it were used. Whenever I went into the laboratory it seemed that the Mary Pickford picture was running. It was fascinating for me to have Farnsworth point out the various elements of betterment in the detail of the image since the last time I had seen it.

With the improved picture the faults of transmission could be more closely analyzed. Study of the clearer reproduction showed that much of the blurriness that clouded it was due to a double image, which created the appearance of a sort of shadowed reflection. In addition to the double effect there was a black splotch, or cloud, down through the center of the picture as though someone had taken a dirty finger and smudged it from top to bottom. These were two fundamental faults in the image which, if not corrected, would prevent television from ever having entertainment value. They were matters of major concern and, as time proved, became controlling factors in the development.

Farnsworth was quick to recognize the situation and set about to analyze the picture and find means of correcting the fundamental faults. He discovered that both difficulties were due to the electrical wave form that controlled the scanning of the image at the transmitter and at the receiver end.

The normal electrical wave form is what is known to engineers as a sine wave. That is, the motion of the wave flows up to and away from the crest in exactly the same pattern, forming the normal shape of the wave as is indicated below. If it is understood that the length of the sine wave is measured in time, it is apparent that in the steep portion of the wave the distance covered in the same interval of time is greater than on the flatter portion; therefore, in this section the motion is accelerated. Since this wave form controls the lateral scanning of the image, the rise in the wave operates the scanning from right to left, and the fall in the wave regulates the return scanning.

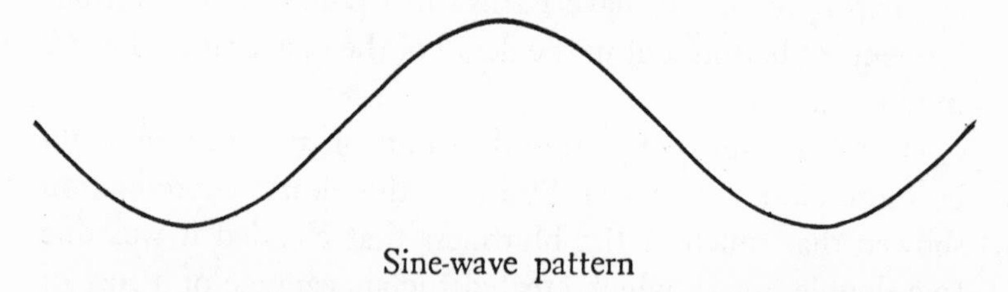

Sine-wave pattern

Actually the sine-wave scanning motion proceeded across the image a good deal as a pen would move back and forth across a sheet of paper in writing exercises, giving a barrel-like pattern. It was found that this sine-wave scanning back and forth caused the double-image effect, the front bulge of the barrel being the main reproduction and the back of the barrel being the blurring shadow. The front image represented the pickup from right to left and the second image or "ghost" was the return line, or left-to-right scanning.

Also it was found that the accelerated motion in the scanning by the sine wave at the steep portion of the course caused the black blur down through the center of the picture. The steep slope of the wave represented the scanning at the center of the image. Since the scanning was faster during this part of the cycle, fewer electrons were released from the surface and the generated current was therefore weaker. This resulted in the dark smudge across the center of the picture, which represented a composite of all the steep parts of the many sine waves that accomplished the complete scanning of the image.

This effect was anticipated by Farnsworth in his first patent application. With his amazing ability to visualize the paths of electrons and their effects, he had in the early theoretical conception of his television scheme recognized that the ordinary sine-wave scanning pulses would not be adequate for his purpose. In this early document he stated that he would use a straight-line, or saw-tooth, wave form for scanning purposes. Such a wave form was much easier to describe in a patent application than it was to produce by his generating panels. However, faced with the reality of a blur across the center of the picture, the production of straight-line scanning, which would give even values all the way across the field, offered the only means of correcting the blur through the middle of the image.

Saw-tooth wave pattern

The saw-tooth form of wave, with a ratio of at least ten to one between the length of the upsweep of the wave motion and that of the downsweep, was necessary to eliminate the double image. In other words, the ratio in time between the scanning from right to left and the scanning from left to right was ten to one. This effectively blanked out the ghost image due to the return-line scanning.

The solution of the problem was first worked out mathematically; then months and months of patient effort were devoted to the development of the saw-tooth wave form. This method is now in universal use as a generating pulse to control television scanning. Along with the initial conception of his electronic television scheme, this achievement ranks among Farnsworth's greatest contributions to television. By the generation of the saw-tooth wave form he had succeeded in clarifying the television image and freeing it from blur and smudge.

The pursuit of this elusive wave form was one most fascinating to the observers of the progress Farnsworth was making in his television image. The results obtained were easily discernible from week to week. In the beginning the efforts seemed to be hopeless, but gradually the smear down the center of the picture disappeared, the ghost image faded out, the picture field cleared, and the image became sharper.

During the period of Phil's intensive work on this problem he had two scanning chassis built. One illustrated the sine-wave scanning and the other the saw-tooth wave he was striving to perfect. The problem he was struggling with was clearly pointed out to me on one of my visits to the laboratory in the early part of the development.

Stepping up to one of the panels Phil turned on the sine-wave scanner. "See that ghost image there?" he asked.

It was unmistakable. There was a second image in the background whose brilliance in comparison with the primary picture was reminiscent of the relative light values of the primary and secondary rainbows in the sky.

"If you'll look closely," Phil said, "you'll see that the scanning lines have sort of a circular path. The ghost image is in the return lines. We've got to get rid of it by straight-line scanning and by cutting down the time of the return scanning so that this ghost disappears."

Then he turned on the image on the panel generating the

saw-tooth wave form. The picture was relatively clear, though there was still a shadow of the ghost image.

"We are scanning now on about a seven-to-one basis," Phil explained. "We hope to get the time ratio of the trip across to the trip back down to ten to one. Then I think the ghost will be out entirely."

The next problem to be tackled was the refinement of the picture to enable the television image to compete favorably in sharpness and clarity with photographic reproductions in magazines and motion pictures. About this time Farnsworth had added to his staff three or four qualified engineers, who were of great value to him in carrying out the details of his research program. To one of these was assigned the problem of finding out what refinement would be required to make television comparable to motion pictures.

Farnsworth hit upon the scheme of taking pictures of a photograph through screens of varying mesh. Facing page 193 are the results of this test. It will be noted that the picture taken through a 50-line screen is very blurred indeed. This, however, was somewhat better than could be accomplished by the mechanical methods used by Baird and Jenkins, who were then operating on 40-line detail pictures. The clarity increased with the number of lines in the mesh of the screen through which the picture was photographed. It will be observed that a 400-line mesh gives a reproduction somewhat comparable to the original. With this chart before him, Farnsworth and his staff set somewhere around 400-line detail as the goal for television scanning. It was decided further that because of the 60-cycle current it would be advantageous to scan the pictures at the rate of thirty frames per second instead of the twenty-four used in the projection of motion pictures.

Newspaper reproduction of pictures is achieved by the use of a screen of fine mesh in the etching process. In television it is well to visualize the scanned image in the same way. To scan a picture of 400-line detail the picture must be broken up into 160,000 units—a number which represents the square of 400. If the pictures are to be scanned at the rate of thirty pictures per second, we must multiply 160,000 by 30, which gives 4,800,000 units per second that must be recorded electrically, transmitted through the ether, and restored at the receiver end in their precise order to achieve the transmission of a satisfactory television picture. To accomplish this without blurring or distorting the image, the synchronization must be tuned to four-millionths of a second. The speed involved is astronomical. This is why the Farnsworth conception of electrical television, as against any mechanical method to produce the same results, is so fundamental.

The controlling factor in increasing the clarity of the image was the sensitivity of the photoelectric plate within the dissector tube, because it is obvious that the finer the units into which the focused image is broken up for transmission, the smaller will be the electrical emissions from the individual scanning areas. For instance, the electrical emissions from 1/2,500 of an image, the unit in scanning a 50-line picture, would be far stronger than the emissions from 1/160,000, which would represent the scanning units of a 400-line picture.

Tests proved that potassium hydride was inadequate as a cathode surface. A beginning had been made with the use of cesium oxide, and while it showed promise, Phil hoped to find a surface that would be better. Russell Varian, who has since achieved fame as the inventor of the Klystron tube, had just joined the laboratory staff. Phil gave him the assignment of finding the most sensitive photoelectric surface. Tests were made of every conceivable material for its photoelectric properties. It was a seemingly endless effort. After testing hundreds of elements and compounds nothing more sensitive was found than the cesium oxide.

Great skill was required on Gardner's part to get a uniformly even coating of cesium oxide—theoretically not much more than one molecule thick—over the emitting surface within the dissector tube. Care had to be exercised that none of the compound was scattered over the glass or other elements within the cell. It was not an easy assignment for a novice handicapped with the meager facilities of the Green Street laboratory, but study and experiment brought some improvement. Purity of the elements involved was a prime consideration.

Scientific laboratories have developed a means of determining the sensitivity of various photoelectric compounds by measuring the intake of light and the output of electrical current as so many microamperes per lumen, the microampere being one millionth of an ampere, and the lumen being the light of one candle at one foot distance. Farnsworth was successful in securing dissector tubes with surfaces giving around twenty microamperes per lumen in his best products. This was adequate for the time being, though the maximum sensitivity appeared in the infrared or heat portion of the spectrum. The tubes were not so sensitive to the usable cold lights.

The percentage of waste in tubes was very high because of the necessity of maximum sensitivity. Farnsworth and his staff were discouraged and baffled by their inability to standardize the production of dissector tubes. However, as the years rolled on and other laboratories reported the results of their experiments in this direction, it became apparent that the percentage of good tubes produced in the Farnsworth laboratories was somewhat higher than that in some other more prominent American and foreign laboratories.[1]

The pursuit of sensitivity in the dissector tube led Farnsworth into many avenues of research, because he recognized it was the bottleneck of television. Among other things, he sought ways and means of increasing the electrical signal generated within the dissector tube before it left the envelope. The research in this followed two major lines: one was that of storing up the electrons on the individual points of the cathode surface during the thirtieth of a second consumed in scanning each frame; the other was the use of secondary emission properties of metals to build the signal before it left the tube. These two methods will be discussed in detail in a later chapter.

[1] The Farnsworth Television & Radio Corporation laboratory has installed an air-conditioned room in which to build these tubes; this has greatly reduced the wastage.

(((9)))

First Demonstration to Backers

Work at the Farnsworth laboratories had proceeded quietly. No effort had been made to publicize what was going on. In fact, it seemed to Farnsworth and his backers that it was best to say little about what was being attempted until some concrete results were obtained. Work went on in this manner until Farnsworth was able to transmit a motion picture with detail of somewhere between 100 and 150 lines at a frequency of 30 pictures per second. This gave a creditable television demonstration if great care was taken in the selection of the subject matter.

Pictures with too great density of color were difficult to transmit. Where there was too sharp contrast the results were not satisfactory. In the latter case the apparatus had a tendency to "overload." This is an engineering term to indicate that too great variations of light and shade are not easily handled. However, the results obtained with carefully selected film were astonishingly good.

Phil had concentrated so intensely on his work that it was quite a shock to him to realize that two years had passed and more than double the money agreed to in the beginning had been spent. He was grateful for the consideration of the backers, but felt uneasy because he had so greatly underestimated the time and money required.

Late in the summer of 1928 Phil talked the matter over with me, and it was determined that his picture was good enough for formal showing to the sponsors. Mr. McCargar was away on vacation, so Phil called Mr. Bishop and invited him and his associates to come to the laboratory.

Phil put on an excellent demonstration. It was the first television picture they had seen, and all seemed pleased and happy that their flier had turned out so well. Mr. Bishop, speaking for the backers, expressed their pleasure. He went on to say that up to this time about $60,000 had been spent on the laboratory work, that it was far in excess of the commitments made originally, but that he felt the results obtained justified the outlay.

At the completion of the demonstration we all gathered in the office, where Mr. Bishop became quite formal and said, "Phil, I congratulate you on the success of your accomplishment. You have done what you set out to do and we feel that thus far you have fulfilled your contract. Speaking for the trustees, we feel that we, too, have lived up to our contract. Now it is a question of what shall be done with this development. It is my opinion that it will take a pile of money as high as Telegraph Hill to carry this thing on to a successful conclusion, and I feel that we should take immediate steps to place it with one of the large electrical companies where there will be adequate facilities for its development. I think it is time to incorporate this undertaking and then take steps to dispose of it in some way or other."

While Farnsworth did not answer Mr. Bishop at the time, I knew that he was chagrined at the prospect of selling, and I said that I thought it would be well to postpone any action until

Mr. McCargar's return. Phil felt as I did, that the results were yet far from perfect and that it would be difficult to interest any of the large electrical companies at this juncture. However, things went along as before for several months, even after Mr. McCargar's return.

About this time some articles appeared in technical journals about the Farnsworth television developments, and these were followed by varied comments on the part of engineers and scientists. The general reaction was one of discouragement. One article in particular pointed out at some length the insuperable difficulties in the path of the successful accomplishment of the transmission of pictures by purely electronic means. The Jenkins laboratories and Baird of England were receiving considerable publicity on their scanning-disk development. The work of Dr. Ives of the Bell Laboratories, and Dr. Alexanderson's demonstration at Schenectady, New York, had also created widespread interest. All stressed the use of the mechanical scanning disk. Electrical engineers and scientists generally were not yet ready to accept Farnsworth's revolutionary theory.

Among the technical help at the Green Street laboratory was Harry Lubcke, a recent graduate of the University of California. He was an able engineer and a good mathematician. Lubcke had an unusually keen publicity sense for one in scientific pursuits and got Farnsworth's permission to write some articles for the technical journals. These further publicized the work going on at the Farnsworth laboratories.

With the increase in the staff, the laboratory expenditures mounted accordingly, and the backers were beginning to feel concerned as to where it was all leading. Finally it was determined to go ahead with the incorporation of the venture.

On March 27, 1929, the venture was incorporated as Television Laboratories, Inc., under the laws of the state of California, with an authorized capitalization of 20,000 shares. Ten thousand shares were issued to the trustees and the original partnership of Everson, Farnsworth and Gorrell.

This was the era of wonderful nonsense in the financial world, and there were many promotional proposals made to Mr. Farnsworth and his associates in an endeavor to bring the public into this speculative venture. Television was a magic word, and to any promotionally minded person the viewing of a television picture fired his imagination with visions of infinite profits. It is to the great credit of Mr. McCargar, in particular, and the others associated in the company, that at no time were they influenced by such proposals. We all felt that the undertaking was highly speculative. We felt we should pocket the loss if we failed, and be in a position to garner the profits if the venture should be successful.

By this time some of the local newspapers had inquired into what was going on at the Green Street laboratories, and several stories were published regarding the television developments. This increased the ever widening interest in the glamorous venture among San Francisco people. The coupling of conservative Mr. Fagan's name and his associates with the genius of the young Utah inventor was an intriguing story.

The expenditures of the laboratory were conservatively handled, and the backers at all times felt they were getting full value for the money being expended. They recognized that there must of necessity be a great deal of money spent in the futile efforts of experiments. The very nature of the venture presupposed that many trials and failures would have to be chalked up to arrive at success in an untried field. In his enthusiasm Phil undoubtedly made many expensive experiments that could have been avoided by a more judicial approach to the problem in hand. He was young, and the generosity of the backers in giving him a free hand did not tend to increase his appreciation of the value of money.

It is difficult to give any conception of the period of tedious development that followed Farnsworth's first successful demonstration to his backers. It was patient, day-in-and-day-out construction, tearing down, testing and revising, month after month with no appreciable daily improvement but with distinctly discernible advance over an extended period.

(((10)))

Television Broadcasting and Transmission over Telephone Wires

After the successful accomplishment of picture transmission by wire, the problem of the wave bands required for television came to the fore as a major obstacle to be overcome. The electrical engineering profession as a whole was quietly skeptical of television because of the breadth of wave band needed. It became apparent that in order to transmit a picture of sufficient detail and clarity to give entertainment value, the wide range of variations involved would require space in the radio spectrum comparable to that used by a hundred commercial radio stations for sound.

At this juncture in radio development there were no such channels available for use in the new art. While it was recognized that they might become available if the ultrashort waves could be made usable, no tubes or circuits had yet been devised to enable engineers to put them to practical use.

This matter was of major concern to Farnsworth. He recognized it as something that must be overcome before television could possibly be a commercial reality. If there were no channels available it would be automatically barred from the air. Phil determined to do something about it. For some time a mathematical conception had been taking form in his mind. To work it out he absented himself almost completely from the laboratory for a period of three months, spending the time in his study at home working with his wife over a scheme for narrowing the wave band. His theory was that the wave-band requirements for television could be successfully narrowed and the picture signal funneled into a portion of the spectrum similar to that required by sound broadcasting. This theory was based on abstruse mathematical calculations.

One morning Farnsworth asked me to meet him at the laboratory. When I arrived he showed me two sheets of engineering drawing paper with curves plotting the mathematical base of his wave-band scheme. Later I learned that these two simple curves represented the solution of literally thousands of mathematical equations which he and Pem had worked out together in the three months' period at home.

Farnsworth threw the two sheets on the desk and said, "George, there is your narrowed wave band." He asked me not to tell any of the boys in the laboratory about it until he had made a "breadboard" setup to demonstrate it.

He got one of the boys to help him prepare the equipment, but did not disclose what he had in mind. Within two hours he had it completed and hooked into the circuit between the transmitter and the receiver. Then he called in all of the laboratory staff and told them what he was attempting to do, and the test was made over the wired lines. The results were exactly as predicted. Lippincott was called in consultation and approved the work. Later the matter was submitted to some consulting engineers for further checking; they also could find nothing wrong with the mathematical calculations.

A short-wave engineer was put to work to prove it out over the air. This proof was never achieved, though a great amount of effort was expended on it over many tedious months of testing and experiment. It was one of the big failures in Farnsworth's development work, and it later became a source of considerable embarrassment. While it has not yet been proved by anyone to my knowledge that there is an error in the mathematical calculations, the thing did not work, and it has never been made to work. No one ever approached a problem more earnestly or with more sincerity, or spent more intensive effort to find a successful conclusion, than did Farnsworth. The theory worked in a test demonstration over wires, but somewhere along the line Phil felt that there was a fundamental fallacy which prevented his getting results in sending the signal through the ether. It was a bitter disappointment to him, both from the practical standpoint and from the standpoint that his genius for mathematics had failed him.

Farnsworth never sought the solution of a problem exclusively in one direction. He always had alternate schemes in his mind if one should fail. He had a happy faculty of forgetting failure and looking about for new ways to succeed. As one alternative he had studied the possibility of its transmission over telephone wires for the short distance between the substation central and the home phone. He felt that the attenuation of the signal over the ordinary wire would not be sufficiently great to prevent its being sent from the central's location into homes in the neighborhood.

To test this theory a room was leased on one of the upper floors of the Hobart Building in San Francisco, about a mile distant from the lab, which necessitated the signal's passing through two telephone exchanges. Test receiving apparatus was installed and the telephone connection made with 202

Green Street. Telephone Company engineers seemed amazed that such an attempt should be made.

Several months were spent in working out the test, but the Farnsworth laboratory did not have at its disposal the experience or the facilities to carry out such an ambitious experiment. While the results were not satisfactory, a picture was actually transmitted. At first it was complicated with a great deal of distortion and "noise." Later the image cleared somewhat, but the expense involved precluded carrying the work further. This was the first time that an electronic television picture was sent from one location to another through the ordinary commercial telephone exchanges.

Shortly after this a setup was made in the Merchants Exchange Building, also about a mile distant from 202 Green Street, but in plain sight of the roof of the laboratory. A low-power radio transmitter was installed with a directional antenna, and a picture was sent through the ether from the roof of the laboratory to the Merchants Exchange Building. To my knowledge this was the first picture ever transmitted through the air by electronic television.

Earle Ennis, a special writer for the *San Francisco Chronicle*, became interested in the Farnsworth developments. Besides writing his popular "Smoke Rings" as a daily feature, he had long had an interest in radio as an amateur. He and his son had quite an elaborate "ham" transmitter in the attic of his Berkeley home. The Farnsworth developments fascinated him. As a result of this he wrote a delightfully clear, popular article about Phil and his television system. The account was featured as a front-page story in the *Chronicle*. It created widespread local interest in the laboratories and was the beginning of considerable notice of Farnsworth's developments by the local and national press.

The Ennis article made us aware for the first time of the possible commercial repercussions of our developments on the radio industry in general. A day or so after his story appeared we were called on by a committee of the San Francisco Radio Dealers Association, who indicated that publicity on television had a very definite effect on sales of home radio sets. The committee also called on the newspapers. This was the beginning of a long series of episodes where our efforts to promote public interest in television seemed to run contrary to the sales promotion campaigns of radio manufacturers and dealers.

It is not surprising that after Earle Ennis' article and other news stories, the sponsors of the Farnsworth company were overrun with various promoters trying to get a foothold in what they deemed a fabulous new art with unlimited financial possibilities. Some of the promoters brought what seemed to be legitimate interest into the picture. Representatives of many of the larger corporations of the country came to San Francisco to see the results obtained at the Farnsworth laboratories.

One of the most interesting episodes resulting from the efforts of volunteer promoters was the visit of Mary Pickford. One of the volunteers had secured the permission of the Farnsworth management to arrange for a demonstration for Miss Pickford, Douglas Fairbanks and his brother, and Joe Schenck, the movie producer. The party arrived at 9:30 on the Lark from Los Angeles and went directly to the Mark Hopkins Hotel to await a call from the laboratory.

The preceding day Farnsworth had shown the promoter an excellent demonstration of television. However, Farnsworth, as was always the case, was not fully satisfied with the picture. He felt he had just made a discovery that would easily make a vast improvement. Consequently in the afternoon he and his staff set to work to make the necessary changes. The result was that he put the whole system out of commission, and after having several engineers work all night, there was no picture to be shown the motion picture people when they arrived.

Phil and the boys worked feverishly through the day, with the promoter frantically calling the laboratory every hour or so in an effort to get a demonstration for the important group from Hollywood who were cooling their heels at the hotel. Finally at five o'clock in the afternoon the laboratory staff had a picture of sorts, but nothing comparable to the results shown the day before.

Miss Pickford and the others were brought down to the laboratory for the showing. The loop of film of Miss Pickford combing her hair was used. The long-count episode of the Dempsey-Tunney fight was also shown, but they were not able to demonstrate transmission from life. To one who had never seen a television picture before, any demonstration which showed a picture was a revelation. However, it was a bitter disappointment to everyone, excepting Miss Pickford's party, that they were shown such a miserable image.

Farnsworth was particularly chagrined and disappointed. He had the awe and adoration of a picture star common to all those of his age. To have failed to put his best foot forward was a tough experience for him. It will be to the everlasting credit of Miss Pickford, so far as he is concerned, that on the train home she penned a note congratulating Farnsworth on his results. But his failure still rankled. This story, like many others, had a happy ending. Several years later, after the laboratories had been moved to Philadelphia, Miss Pickford was playing an engagement at a theater there. Farnsworth was able to invite her to his new laboratories and show her a first-rate studio demonstration of television transmission from life.

As Farnsworth's developments progressed, work on mechanical disk television had also made rapid strides. Baird of London had received a great deal of publicity on the work he was doing. The Baird Company had started an experimental broadcast station at the Crystal Palace in London and had sold some moderate-priced mechanical scanning-disk receiver sets to the British public. Dr. Alexanderson of the General Electric Laboratories, Dr. Ives of Bell Laboratories, and Jenkins of Baltimore had all received continuing notices in the public press. The Bell Laboratories had set up a simple demonstration in New York. Dr. Alexanderson had shown to the press the results of his work, and Jenkins was endeavoring to promote his developments commercially. The Jenkins activities particularly tended to bring television very much to the fore. A financing of major proportions had taken place, and its attendant publicity had raised the hopes of the general public regarding the possibility that commercial television would soon be introduced. The public became greatly intrigued by the new art and every sign of progress was received with widening interest.

As a result the Federal Communications Commission, then known as the Federal Radio Commission, took cognizance of the experiments in television and scheduled an informal hearing on the possible assignment of wave bands for its use. The hearing was set for December 1930. Farnsworth and McCargar attended.

This marked Phil's first important public appearance. He made a very favorable impression as an authority on television. In addition he highlighted his testimony by disclosing his ill-fated plans for narrowing the wave band for television broadcasting. Phil's advisers were aware of the hazards involved in such a startling announcement and had taken the precaution of having Farnsworth's mathematics checked and approved by the Naval Research Laboratories of Anacosta, D.C., before allowing him to make any public statement regarding his new discovery. It was an unfortunate disclosure, because later, when his findings were found to be inoperative, despite their checking and approval by experienced outside engineers, it did Farnsworth a great deal of harm among the radio engineers, some of whom even today speak slightingly of Farnsworth because of this failure.

His youth and the startling nature of his inventions brought Farnsworth into the public eye to an extent that was somewhat embarrassing to him and his sponsors. However, while the publicity was generally not advantageous, it did bring the accomplishments of the laboratory to the attention of the engineers and executives of companies of importance in the electrical and radio business.

One morning when I dropped in at the lab I found Farnsworth in a most jubilant mood because he had just received word that Dr. Vladimir Zworykin, of Westinghouse and R.C.A., was shortly to visit his laboratories.

Farnsworth had a high regard for the scientific abilities of Dr. Zworykin, who, through his research and writings, had achieved considerable reputation at the Westinghouse laboratories. In his work there Zworykin had had some success in the transmission of pictures from film by the use of an oscillating mirror at the transmitter and the cathode-ray tube at the receiver. At this time the research work of R.C.A. was being concentrated at Camden, New Jersey, and Dr. Zworykin had been transferred from the Westinghouse laboratories to the new R.C.A. laboratories at Camden.

Farnsworth had knowledge of Zworykin's work and told me that there was no engineer in the country he would rather have view his results. The coming of the scientist was therefore awaited with much interest. Phil felt that here at last would be someone from the outside who understood the language he spoke and who had the proper appreciation of what he had accomplished.

Upon his arrival Dr. Zworykin was gracious in his praise of Farnsworth's results and seemed tremendously impressed with what he saw. He was shown in detail all that had been accomplished. Sitting at Farnsworth's desk on the first day after his arrival, in the presence of Mr. McCargar, Mr. Lippincott, and me, he paid Farnsworth high tribute by picking up the Farnsworth dissector tube and saying, "This is a beautiful instrument. I wish that I might have invented it." This, of course, made Phil very happy. It was the type of recognition that he felt he must have if his efforts were finally to succeed in a substantial way.

Zworykin was astonished at the success Gardner had had in sealing an optically clear disk of pyrex glass into the end of the dissector tube. He asked Farnsworth how they came to find out that such a thing could be done. Farnsworth replied that it was necessary to have a tube of that kind, so they went ahead and made it. Dr. Zworykin said he had consulted with the best scientific glass blowers in the laboratories of both Westinghouse and R.C.A. and they assured him repeatedly that a seal of that nature could not be made. He went on to say that he would be greatly in Farnsworth's debt if he would ask Gardner to let him see how he accomplished the sealing process. This was later arranged, and Dr. Zworykin saw Cliff Gardner make the dissector tube.

Dr. Zworykin spent several days at the laboratory. He and Phil were constantly together checking and discussing the many phases of the television development. The visit gave Phil new confidence and enthusiasm for his work.

In the history of television the names of Zworykin and Farnsworth lead all the rest. During the time that Farnsworth was perfecting the first model of his electrical system, Dr. Zworykin was working on his oscillating mirror system in the Westinghouse laboratories and later at the R.C.A. laboratories.

Previous to this time Zworykin sold a conception of electronic television to the Westinghouse Company. This invention was the forerunner of the system that was later developed by Dr. Zworykin at the R.C.A. laboratories. The heart of this system was the Zworykin iconoscope, which scanned the image by electronic means, though in a somewhat different way from that used in Farnsworth's dissector tube.

The Farnsworth disclosures helped convince Zworykin of the practicability of the purely electronic method of television transmission and reception and led to his devoting his entire future effort to perfecting his own invention centering around the iconoscope.

((11))

Farnsworth Moves to Philadelphia

WHEN DR. ZWORYKIN was invited to visit our laboratories, Farnsworth and all of us recognized that we were courting competition of the keenest sort. Farnsworth knew enough through published reports and through interferences he had encountered in the Patent Office to appreciate that a great risk was being taken by making complete disclosures to the eminent scientist. It was not fear that Farnsworth's ideas would be stolen, but that it would spur Dr. Zworykin on to intensive work that would be highly competitive. Yet, we reasoned, if television was ever to become a commercial reality, it was felt that R.C.A., one of the leaders of the radio industry, must have a hand in it.

On the whole, the decision to show Dr. Zworykin everything was a wise one. Though it did result in more active work on the part of the Camden laboratories, it also brought a powerful ally into the field of television research and development.

While the visits of Zworykin and others to the laboratories were very exciting for the young inventor, he did not lose sight of the importance of intensive effort in perfecting his invention. He recognized that he was still a long way from commercial television. The question of the sensitivity of the dissector tube was the major problem always dogging his research activities. For the successful transmission of pictures, infinitely small electrical units must be detected and stepped up. He was constantly seeking new avenues of experiment to bring about improvement in sensitivity and in amplification.

Ever since Lee De Forest invented the screen grid tube, electrical engineers have been plagued by the phenomenon they call "secondary emission." It was learned that within a vacuum tube every metal has in suspension on its surface free electrons that are released when they are bombarded by other electrons moving at high speed. In the operation of amplifier tubes random electrons often released "secondaries" that were sources of "noise" and other annoying reactions within the radio circuits.

In canvassing the possibilities for greater and more efficient amplification, Phil hit upon the idea that this disturbing phenomenon might be put to practical use. Time and again he wrestled with the idea, only to set it aside and return to it later. If it could be harnessed, he believed, almost unbelievable amplification could be attained. The catch was to devise a harness that would curb and put the fractious electrons to work. He believed that if he could establish suitable circuits he could, by bombarding properly charged plates with free electrons, set up a multiplication of electrical energy that would proceed by geometrical progression.

It was known that if a volley of free electrons was fired at a photo-emissive surface, each bombarding electron would release from two to six secondary electrons, depending on the metal used, thus strengthening the flow of electrons by that amount of multiplication. Farnsworth conceived that if he could oppose two plates within a vacuum tube, and bat electrons back and forth from one surface to the other under proper control, he could get almost infinite multiplication in a split fraction of a second. In other words, each original electron, upon hitting the first surface, would have a "litter of from two to seven puppies." These would be released at high speed and proceed to the opposite plate with the parent electron and each one of them, upon impact, would free another litter of electron puppies. Since the speed of electrons is equal to the speed of light, it was calculated theoretically that once this process was set in motion, within a second's duration the multiplication of power would produce as much electrical current as is now available in all of the electrical plants in the world. Naturally no such power could be generated within a small vacuum tube without destroying everything, including the tube itself; however, under proper control Phil believed it could be used as a most efficient means of radio amplification and have far-reaching effects in the entire realm of electronics.

The more Phil explored the possibilities of this conception, the greater became his enthusiasm for it as a solution to his amplifier problems. The method of stepping up the current that it offered seemed so straightforward and right that he determined to tackle it as a major research problem. To accomplish his ends two lines of attack seemed necessary: first, to determine what metal was most efficient for the purpose, and second, to design a tube that would harness the electrons for practical use.

Following the first line of attack, he knew that there were great variations in the secondary-emission properties of metals. Some metals released only one or two secondaries. Others released more. To achieve maximum results he wanted to find the surface that would release the largest number of electrons. Russell Varian, a well-equipped electrochemist, was set to work to carry out original research on the subject. The problem was somewhat similar to the one encountered in making the photoelectric surfaces for the dissector tube. Over a period of two years Varian, a thorough and resourceful engineer, tried everything from common table salt to platinum. In the course of his research he turned up many helpful suggestions quite outside his direct line of experiment. It was finally determined that cesium oxide on a coating of silver was the most effective surface for this purpose, as well as in the dissector tube.

While Varian was working on the surfaces, Phil was devoting much time to a design for a tube that would use the "secondaries" to advantage in amplification. It was a baffling and tricky problem. Farnsworth finally disclosed to Lippincott a strange-looking device which later became known as the "pistol tube." The name was derived from the fact that the tube was shaped somewhat like the ordinary pistol. It was the first successful effort Phil had made in his endeavors to control and put into service the electrons produced by the secondary-emission phenomenon.

As was often the case in his research work, Phil abandoned the project temporarily and turned to other things. He was a great believer in letting ideas of this kind ripen for a while, rather than artificially forcing the issue. As we shall see later, the conception was not given up, but was simply held in abeyance to await the gestation of new thoughts regarding it.

Although the pistol tube did not help Farnsworth in the solution of his amplifier problems at this time, it was the forerunner of later developments which became known as the electron multiplier principle used in the creation of the Farnsworth multipactor tubes.

Somewhere in his early reading on scientific subjects, Farnsworth had been greatly impressed by the fact that De Forest had won his most important patent suit by the introduction of a laboratory notebook showing the conception date of the idea in controversy. This made Phil extremely conscious of the value of full and accurate laboratory notes and caused him to set up an efficient laboratory notebook system for himself and his associates. During the first three or four years of his work the more important records on his inventions were transcribed from the original books to bound leather-covered volumes. Most of the early notes and sketches are in ink in Phil's nervously precise handwriting. These volumes, and twenty or thirty volumes of original laboratory notes, constitute the background for the Farnsworth patent structure.

In addition to having a proper respect for well-kept laboratory notes, Farnsworth had an unusual flair for sensing valuable patentable material. As a result, very early in his career he and his patent attorney had laid a broad foundation for a comprehensive patent structure covering the field of television, and in addition, much pioneering in the field of general radio and electronics.

The months following the demonstration to the financial backers were anxious ones for Farnsworth. His interest was divided between the feverish activities of improving the picture and the uncertainty as to what the sponsors would do with their interest. Phil did not want to sell out control lock, stock, and barrel. He felt that it was premature. He was therefore greatly encouraged when in the spring of 1931 Mr. McCargar and I succeeded in enlisting the interest of the Philco Company in the Farnsworth television developments.

Walter Holland, vice-president in charge of engineering for Philco, came to San Francisco to inquire into the Farnsworth developments after he had received a favorable report from Larry Gubb, the company's sales manager. Mr. Holland was so impressed that he remained in San Francisco several days to work out with us the first draft of a Farnsworth licensing arrangement to be presented to the Philco board of directors. Following this, McCargar and Farnsworth went to Philadelphia to complete the arrangements.

In June of 1931 the contract was formally entered into with the Philco Company for the licensing of television receiver sets. It was the first substantial recognition by a major radio or electrical company of the Farnsworth system. Naturally Farnsworth was elated by the tie-up with a company which at the time was the largest manufacturer of radios in the country.

As part of the licensing arrangement it was agreed that Farnsworth and his staff were to move the laboratory to Philadelphia in quarters provided by the Philco Company at the Ontario and C Street plant. It was decided that Phil's entire staff, with the exception of the two engineers, would go with him. As a measure of precaution, Mr. McCargar and I determined that the Green Street laboratories should be continued with two or three picked engineers to carry on development work supplementing the operations at the Philco plant. However, a major portion of the laboratory equipment was moved to Philadelphia.

When the time for departure arrived there was great commotion at the laboratory. The packing of the tubes and equipment needed to be done with great care. No one realized the bulk of it all until it was packed in boxes and ready for loading on vans.

The move to Philadelphia made an equally great change for Phil's family, consisting of his wife and their two little boys. Phil and Pem had become much attached to their California home, and it was therefore with much reluctance that they moved out of it for an unknown future in the East.

Upon arrival at the Philco plant Phil found that a penthouse laboratory had been provided on the roof of the manufacturing building. To a group used to the cool San Francisco summers, the change to the stifling July heat of Philadelphia, accentuated by the radiation from a wide expanse of glaring sun on the roof around the penthouse, was quite a hardship. The quarters were much smaller than the Green Street laboratories, and the addition of several men made them somewhat cramped.

At the Philco plant Phil made his first acquaintance with a large manufacturing operation. He was required to fit his budgeting and laboratory requisitions into the routine of factory operations.

Little was heard from Phil regarding the progress being made. Again his complete deficiency as a correspondent became evident. His concentration on his work made it impossible to recognize the necessity of keeping his associates informed. Mr. McCargar and I bore with this failing with good grace, feeling certain that no news was good news and that if Phil got into serious difficulties we would hear from him.

At the end of the first year's operations the Philco executives asked Mr. McCargar and me to come to Philadelphia. The visit was most interesting and pleasant. We found Phil's work surrounded with the greatest secrecy. So strict were the safeguards that Mr. McCargar was prompted to remark that he doubted if the Philco executives talked even to themselves. No one except Mr. Holland, the vice-president in charge of engineering, Mr. Grimditch, under whose department the laboratory functioned, and Mr. Skinner, the president, were allowed to visit Phil's sanctum. This complete isolation had no doubt contributed to Phil's reticence in writing to us in San Francisco. I am inclined to think that the policy of secrecy was designed more to prevent annoyance and interruption by the curious than to prevent competing interests from learning what was going on.

At the Philco plant Mr. Holland and Mr. Skinner had taken an understanding attitude toward the uncertainty and unpredictability of the research program, and Phil had found the atmosphere most congenial. Here for the first time he met face to face the practical commercial problems involved in the introduction of the new art of television. It was determined that experiments should be made with actual broadcasting over the air. This made it necessary to apply to the Federal Communications Commission for an experimental broadcast license. After several months of delay, the license was granted to the Philco Company. This added new aspects to Phil's problem and necessitated broadening the field of research.

Here also Phil had his first chance to observe the work of a rival inventor. Through their receiver set in the penthouse he and his staff often picked up the Zworykin pictures being broadcast from the R.C.A. laboratories at Camden. At first they made Phil uneasy and nervous, but later he got in his stride and took advantage of Zworykin's broadcast to study it for checking and evaluating his own developments. When Mr. McCargar and I arrived in Philadelphia, the reception of the R.C.A. pictures was among the first things Phil discussed with us. He praised the Zworykin results, and they seemed to act as a stimulant to his own work.

During the second year at the Philco plant it became apparent that Farnsworth's aim in establishing a broad patent structure through advance research was not identical with the production program of Philco. So at the end of the second year the Philco Company established its own laboratories and, after careful survey of the situation by Mr. McCargar, Farnsworth, and me, it was determined to establish our own laboratory in the Philadelphia area. It seemed advantageous to have our main operating base close to the center of the radio industry.

This was in the summer of 1934, in the midst of the depression years. Phil agreed to a budget of $1,500 a month as a start, since that seemed to be all we could afford in addition to the rapidly mounting costs of maintaining the necessary patent work.

Quarters were leased in a quiet section of Chestnut Hill, and Phil and his staff settled down to work again. Immediately expenses began to mount, and almost before we were aware of it the monthly budget of the Philadelphia plant was running around the $5,000 mark. In spite of all the efforts of Mr. McCargar and me it seemed impossible to cut down the operations. There seemed no other course than to meet the expanding costs by broadening the interest in the company. With the consent of the Corporation Commission, small blocks of stock were sold privately from time to time to meet this increased cost of operation.

(((12)))

Multipactor Tubes Demonstrated

At the Philco plant Farnsworth again attacked the problem of harnessing the secondary-emission phenomenon for use in his amplifiers. After his initial attempts in the San Francisco laboratory, which resulted in the pistol tube, he allowed this idea to remain dormant over a period of a couple of years.

The urgent need of more efficient amplifiers drove him to another attempt at amplification by the use of the electron multiplier idea. He was convinced that the principle would work if he could design the right tube structure. After much experimentation and mathematical calculation he constructed a cylindrical tube sealed at each end with two disks of metal coated with cesium opposing each other at equal distances from the ends. He placed this tube in a controlled magnetic field and turned on the current.

The thing worked! Farnsworth was truly astonished at the results obtained.

Each of the cathode plates operated cold. When the tube was in operation there was no sign that it was alive and functioning, yet it produced amazing amplification of power, apparently coming from nowhere. In an enthusiastic report to Don Lippincott on the success of the experimental tube, Farnsworth said that it "was like getting something for nothing," that the cathodes, operating cold, seemed to pick up amplified power from nowhere and deliver it for use. Because the tube achieved its gain in power by the multiple impact of electrons, it was given the name of the Farnsworth "multipactor."

Some weeks later Farnsworth sent one of these multipactors to the San Francisco laboratory. It was here that Earle Ennis, of the *San Francisco Chronicle*, saw the tube function. Perhaps a clearer understanding of the device may be gleaned from Mr. Ennis' newspaper description of the operation of this tube:

Development of an astonishing new radio-television tube that not only transmits television impulses, but may be used as an amplifier, detector, rectifier and multiplier tube as well, and may make obsolete all known forms of radio tubes, was announced yesterday by the Television Laboratories, Inc. The new tube, according to the laboratories, is the long-sought "cold cathode" tube which has been the goal of laboratories the world over. It is a multiplier of current to an astonishing degree and because of its five-fold function it is of world-wide scientific interest.

To make plain the operation of this child Titan without going into a technical description, it must be understood modern radio tubes are all of the "hot cathode" type. The source of the electrons is a filament. When this filament is heated by passing a current through it, the electrons are "boiled off" by high temperature. The process is scientifically ineffective, as a comparatively large amount of power is required to obtain a small number of electrons.

In "hot cathode" tubes filaments burn out, vacuums vary and because of mechanical difficulties tubes seldom are the same. In other words, little or no standardization is possible.

The "cold cathode" tube, long sought as a solution to radio troubles, is one that has no filament or grid, and so has nothing to burn out. It has no instability due to changes in gas pressure and is, in a sense, perpetual and indestructible. In a "cold cathode" tube there are two plates between which the electrons pass. In the Farnsworth tube they are "bounced" with terrific force on the two plates. In this "bouncing" lies the secret of the tube's amazing new powers.

The "bouncing" process knocks additional electrons from the cathodes, and these in turn are bounced against the electrodes, which, in turn, have more electrons knocked out of them. As each electron may be bounced 100 times to 1/1,000,000th of a second, the tremendously rapid growth of electron progeny may easily be understood.

In the Farnsworth tube a single electron will build up or father 2,000,000,000,000,000,000,000,000,000,000,000,000,000,000,000,000,000,000,000,000,000 electrons, all in the space of 1/1,000,000th of a second. Each of the "children" is a perfect electron, with all the properties and speed of the parent. This, it easily can be seen, produces a terrific multiplication or increase of current, so great in fact, that if it is not drawn from the tube inside of the 1/1,000,000th of a second, the tube electrodes melt.

The multipactor tube was hailed by scientists and engineers as a major development in electronics. The first public demonstration of the tube was given in San Francisco at the plant of Heintz & Kaufman, with a power multipactor which Phil had sent out to Mr. McCargar and me for the purpose. Ralph Heintz placed it in the transmitter circuit of the Globe Wireless station. Members of the Signal Corps of the Army, members of the faculties of Stanford University and the University of California, and other scientists around the Bay region, were invited to witness the showing. When all were gathered, the transmitter was turned on and the tube set into operation. There were no visible signs of anything going on within the tube, or without, yet when Ralph Heintz took an electric light bulb on the end of a stick and placed it within the field of the tube's power, the radiating power lighted the incandescent globe, proving that the multipactor was operating and giving out the power expected of it.

I was standing with Mr. McCargar and Dr. Leonard Fuller,

of the University of California, when Heintz placed the light globe in the field of the multipactor tube. When it lighted up, Mr. McCargar, not being sure of what was expected, turned to Dr. Fuller and asked, "Does it work?" Dr. Fuller, who is more reticent than the average cautious scientist, committed himself to the extent of, "It seems to."

Not until long after this test did I learn through one of the young engineers of Heintz & Kaufman's plant that the demonstration was made without preliminary testings. Ralph Heintz had in his make-up something of the inventor's casualness as to details. While arrangements had been made several days previous for the showing, he did not assign anyone to set it up until the afternoon before the evening demonstration. Since a special antenna had to be constructed, the whole setup was not complete until after the crowd had gathered. So when Ralph turned on the power it was actually the first test that had been made.

Through the circuits set up a message was sent to Farnsworth at Philadelphia, and messages from the San Francisco station were received at different points around the Pacific. Sir Hubert Wilkins picked up the broadcast somewhere in Australian waters.

The new development added substantially to the sensitivity of the dissector tube. Farnsworth and his associates designed a minute multipactor no greater in diameter than a lead pencil for use in the anode finger of the television pickup tube. By this device they were able to increase the strength of the current flowing from the picture scanning manyfold. This made the dissector tube much more practical for picking up studio and outdoor scenes.

The multipactor tube was instantly recognized by Farnsworth and his associates, and by the engineering profession in general, as having broad applications to the whole field of radio. There seemed to be so many fields in which it would be immediately applicable that much emphasis was placed on it as a major line of research. A wide variety of experimental tubes were developed and tested. One tube was made to operate as a complete radio receiver set. Attached to an auditorium loud-speaker it brought in the local Philadelphia stations with blasting intensity. This was probably the first time in the history of radio that a one-tube radio receiving set had effectively brought in a program from the air. Because there was only one tube for amplification, there was a minimum of distortion in the reception, which was astonishing in its clarity and fidelity of reproduction. For the benefit of Mr. McCargar and me this one-tube radio set was duplicated in the San Francisco laboratories in a demonstration set up by Bart Molinari, the brilliant radio technician in charge of the San Francisco operations.

In Philadelphia a sound unit using a multipactor was built for use in a motion-picture projection machine. This sound pickup fed directly into the loud-speaker without intermediate amplifiers and gave excellent fidelity in the reproduction of sound from the film. This was but one more of the possibilities that lay ahead in the adaption of the multipactor principles in the field of electronics.

The tubes had one drawback: they seemed to require too much of the surfaces in operation, so that they showed a tendency to fatigue and deterioration when subjected to life tests. Here again the cathode surfaces were a major problem, and much research was done in an effort to rid the tube of "bugs" before it could be used widely for commercial purposes.

Since then other laboratories in the United States and abroad have done intensive work on the multipactor principle, and it is being gradually adapted to commercial and scientific fields.

(((13)))

Television Show at Franklin Institute

WITH THE establishment of independent laboratories in Philadelphia, Farnsworth and his staff turned their attention to building up a practical demonstration unit for television. Designs were made for a portable transmitter and receiver so that it might be used outside the laboratory. In the summer of 1935 this was completed for demonstration purposes. While both the camera and receiving units operated well enough, they definitely had the appearance of homemade affairs. However, the compactness and mobility of the transmitting units were quite an achievement at the time.

Since his arrival in Philadelphia, Phil had made acquaintance with men in scientific pursuits, and his accomplishments in television were gaining recognition. As a result, the management of the Franklin Institute, after a visit to the laboratories, invited Farnsworth and his staff to make a public showing of television at the Institute for a period of ten days. The invitation was accepted. It was the first time that electronic television had ever been shown for the general public to see.

None of our staff had any knowledge of programs or public exhibitions. They had had a singer or dancer appear before the camera at the laboratory from time to time, and there had been one or two demonstrations to the press, but aside from that their experience was nil. They therefore had to make up the program as they went along.

An improvised studio was set up with the television camera unit on the roof of the Institute. Wire leads were brought into a small lecture-room auditorium with banked seats. The room accommodated about two hundred people. Because of the seating arrangements, all had a good view of the television screen.

The receiving set built by Farnsworth for demonstration purposes was an astonishingly large unit. The received image was approximately 12 x 13 inches. To produce this image a cathode-ray tube fully as large as a ten-gallon water jug had to be used.

The opening of the show was heralded with considerable newspaper publicity. The mayor of the city and representatives of the Institute gave short televised talks. The demonstration was a success from the start. The programs, of fifteen minutes' duration, were carried on from ten o'clock in the morning until closing time in the afternoon. The entertainment consisted largely of such vaudeville talent as could be found. A ventriloquist was one of the most successful acts. Stoeffen and Shields, the tennis stars, gave interesting demonstrations of tennis technique. Had there been a tennis court on the Institute roof, a televised tennis game would have been well handled by the camera. When there was nothing else to show, the camera was pointed at the traffic in the street below and the statue of William Penn on top of the City Hall several blocks distant.

It was during the testing out of the equipment for the Franklin Institute demonstration that, quite by chance, the moon was televised. The engineers had worked late on a midsummer moonlit night. As they were getting ready to leave, one of the boys, remembering the special sensitivity of the dissector tube to the infrared portion of the spectrum, said, "Let's try to get a shot of the moon." It was tried, and a picture of the moon came through beautifully. Next day Phil told a New York newspaper correspondent about it. The reporter made note of it and a press dispatch to a San Francisco newspaper read as follows:

> First recorded use of television in astronomy was announced yesterday in Philadelphia by Philo T. Farnsworth, young San Francisco scientist.
>
> And it was the man in the moon that posed for his first radio snapshot.
>
> Reproduction of the moon's likeness is just another sensational achievement by the young inventor who has been working on his television apparatus since the age of 15.
>
> The picture of the moon was taken, according to Associated Press dispatches, as the ultimate test of the supersensitivity of Farnsworth's television invention.
>
> And a "return performance"—open to the public—is promised by Farnsworth for the benefit of all doubters. This "show" will take place on the "first clear night."

To say that the Franklin Institute demonstrations were nerve-racking for the Farnsworth staff would be putting it mildly. To add to the anxiety attendant upon the demonstration and program material, it was the first time that the laboratories had used the extra-large oscillite tube in the receiving set. Because the tube was so large they were not sure that it might not collapse under the stress of the rather high vacuum with which it was operating. As a measure of prudence to insure continuity of operations there had been brought to the Institute a couple of spare large tubes. None broke while in operation, but the staff found to their astonishment that one of the spares had burst in its box during the night without injury to anyone or anything. This was the first and last time that an oscillite tube collapsed. Farnsworth and the rest gave fervent thanks that this breakage had occurred in a tube not in operation, and in the middle of the night when no one was around.

(((14)))

Farnsworth Visits England and Germany

THROUGH THE activity of the Baird Company, far-reaching developments in television were under way in London. Over a long period of time this company had been broadcasting television by the mechanical scanning-disk method from the Crystal Palace. Also there had been some sales of scanning-disk receiving sets to the public. In the meantime the Marconi Electrical & Musical Industries had been doing extensive work in electronic television developments stemming from the work of Zworykin and his associates of the Camden laboratories of R.C.A. Considerable criticism of the sale of scanning-disk receiving sets to the London public had developed, and the matter received the attention of the British Parliament. As a result a committee was appointed by Parliament to study television developments and to make recommendations as to what should be done with this new art by the government-owned British Broadcasting Company.

As the committee's exhaustive study of television progressed it became apparent that the electronic method of transmission and reception would be approved by the committee and recommended for adoption by the British Broadcasting Company.

Since the Baird Company had built all its developments around the mechanical scanning-disk methods of Dr. Baird, they found themselves in a difficult situation.

Some time previously a young Englishman had called on me in San Francisco and announced his desire and determination to work in the Farnsworth laboratories. Upon finding that Mr. Farnsworth was in Philadelphia, he flew east and made such a nuisance of himself that Phil finally put him to work. Later he went to London and joined the engineering staff of the Baird Company at about the time that Parliament's television committee was making its study. His enthusiasm for Farnsworth's work resulted in the Baird Company cabling Phil, urging him to come to London at once, and to bring his demonstration unit as a preliminary step in negotiating a licensing agreement. We were not inclined to act without having some definite plan worked out beforehand. As a result a Baird representative came over and arrangements were made for Farnsworth and members of his staff to take the demonstration equipment to London in the fall of 1934.

Naturally Phil was greatly exhilarated over this step toward international recognition. There was great hustle and bustle to get everything safely crated for the trip.

Upon arrival Phil became the center of attention of the Baird engineers and their associates. Haste was necessary to prepare for the showing to the parliamentary committee. The camera equipment was installed at the Baird Laboratories in the Crystal Palace in London. A receiving set was placed in an inn at a distance of twenty-five miles for the demonstration to the committee. At the appointed hour the committee arrived and a satisfactory showing was made.

Spark plugs of automobiles have long been recognized as a possible source of interference in television reception. The committee, being aware of this, instructed their chauffeurs to drive slowly past the inn during the demonstration. Fortunately the receiving set had been properly shielded against the short-wave emanations of the spark plugs, and they gave no trouble.

Farnsworth's demonstration was an important factor in the committee's findings, which resulted in a recommendation to Parliament that an appropriation be made for the British Broadcasting Company to establish television service for the London area. The Baird Company and the Marconi E.M.I. were named as the two suppliers for television equipment by the British Broadcasting Company.

The action of Parliament put London at the forefront in television news and developments. As part of their program of publicity the Baird Company did considerable experimental broadcasting of programs. One of these was a fashion show which, among other things, showed the Duchess of Kent purchasing a hat at one of the London stores.

The Baird Company enjoyed a close working arrangement with the Fernseh A.G. of Berlin, a company owned jointly by the German Bosch Company and the Zeiss-Ikon Company. It was headed by Dr. Paul Goerz, who, though not a Nazi, had been appointed as co-ordinator for radio and television for the German Reich. At the suggestion of the Baird Company a representative of Fernseh visited the Farnsworth group in London. Later Phil and his associates went to Berlin. In due course a licensing arrangement was entered into with Fernseh similar to the one that had been consummated with the Baird Company. This provided for the proper introduction of the Farnsworth principles in England and on the continent of Europe.

This trip to Europe was the first of several that Farnsworth took in the interests of television. The experience was a broadening one and gave him the opportunity to meet the leading exponents of television abroad. The work of Fernseh was particularly impressive. The engineers of this company were applying the skill for detailed refinement that is so characteristic of German scientists and engineers. Fernseh's laboratories were located in the plant of the Zeiss-Ikon Company and had the advantage of the tradition for fine workmanship that has made them noted throughout the world for optical and precision instruments.

The Baird Company connection proved most disappointing. At about the time the company was ready to install its transmitting equipment, based on the Farnsworth developments, in the Alexandra Palace for tests by the British Broadcasting Company, a disastrous fire swept the entire Crystal Palace. All of Baird's fine studio and transmitting equipment was destroyed in the flames. Farnsworth was abroad at the time. Upon his return to Philadelphia he showed me, with a wry face, a distorted piece of melted glass with some wire protruding from it. It was all that was left of the dissector tube of the television camera the Baird Company had expected to install for tests during the week in which the fire occurred.

The Fernseh Company, up to the time of negotiating the licensing arrangement with Farnsworth, had concentrated on the scanning-disk method of television. To achieve greater refinement in picture transmission they used motion pictures almost exclusively. One of the most interesting of their developments was a television truck for the transmission of sports and news events. This truck was fitted out with complete equipment for taking sound motion pictures. Also it had facilities for the immediate development of film by a rapid process which produced a developed film for television transmission immediately from the scene of action. This intricate and expensive expedient was resorted to because it was not possible to get good outdoor television pickup with the scanning-disk cameras. At the time this process enabled the Germans to transmit pictures from the scene of action with surprising accuracy and fidelity. Electronic television cameras have since made this resort to intermediate motion picture film unnecessary.

On his last visit to Germany, Farnsworth, because of his international reputation as an inventor, was put under close surveillance by the Nazi police in order to be sure he did not see anything of German scientific developments that they did not want him to. This vigilance became so irksome that Farnsworth threatened to leave Germany if it continued. As a result Dr. Goerz made arrangements with the Nazi Government to discontinue the vigilance. Also it was arranged that Farnsworth was to be shown all of the Fernseh television developments and be privileged to take out of Germany the scientific data to which he was entitled under the Farnsworth contract with the Fernseh Company.

Arrangements also were made for full and complete exchange of scientific data and information of technical developments between the two companies. That arrangement continued without interruption until a state of war with Germany was declared.

As time went on, representatives of both the Fernseh Company and the Baird Company made periodic visits to the Philadelphia laboratories. Dr. Goerz and Dr. Moeller usually came for Fernseh, and Captain West and Mr. Lance for the Baird Company. In turn Phil, or some of our engineers, paid return visits. Phil enjoyed such trips and gained much from them. The visits back and forth, the exchange of research and engineering discoveries and technique, were naturally profitable. Warm personal friendships grew up and flourished. It was an ideal international relationship. This was destroyed by World

War II. Dr. Goerz probably served at the German front. Captain West had served in the radio intelligence service of the British in World War I; he probably acted in some such capacity in World War II.

(((15)))

Patents

FILING A patent on an invention is like planting a crop and waiting for it to germinate and then nursing it through a hazardous season of growth to the harvest.

Phil's first patent application was filed on January 7, 1927. Following that there was a steady flow of new ideas to the Patent Office for protection. In important cases there were anxious months of waiting for the first reports on allowances of claims. Then there were other long stretches of uneasiness when interferences might develop through others claiming prior conception of the same ideas. Phil was aware of these hazards as he followed the progress of his principal invention through the Patent Office.

The interferences not only endangered the very substance of that for which he was striving, but presented serious financial problems that were difficult to meet. The first important one developed when Radio Corporation made an effort to bring the dissector tube under R.C.A. patent domination. This suit struck at the heart of the Farnsworth system. To lose it would have been disastrous. Lippincott and an associate handled the case for us. Two substantial volumes of testimony were taken. It was necessary to hunt up Tolman, who was finally found in a school in Salt Lake City. Expert technical witnesses were called on both sides. The whole procedure called for expensive trips for attorneys to New York, Salt Lake City, Kansas City, Washington, and other places. Winning of the interference was of vital importance to our whole venture; consequently no expense could be spared.

Phil was subjected to days of grueling cross-examinations by R.C.A. attorneys and technical experts. Dr. Zworykin of R.C.A. was put through an equally severe ordeal by Lippincott and his associates. After weeks of careful preparation elaborate briefs were filed by the contenders. The whole proceedings took many months, anxious months for Phil and his backers. Finally all the testimony was in. It had cost the Farnsworth Company about $30,000, at a time when the country was struggling through a major depression. To find money for such litigation fell on Mr. McCargar and me. It was a difficult and exhausting ordeal. After several months the Patent Office Board of Appeals awarded the claims to Farnsworth.

The next recourse was to the Civil Courts. By law R.C.A. was permitted six months to appeal the case. This period was an added time of anxious waiting. If appeal were taken it meant the hazard of a final opinion by a layman judge passing on a highly technical subject. It also meant a difficult and expensive court fight which would have been a heavy additional drain on our finances. We all gave vent to a sigh of relief and a cheer of exultation when on the last day upon which appeal could be taken Lippincott rushed into the laboratory with the news that R.C.A. had abandoned the appeal and the patent was conceded to Farnsworth.

It was a red-letter day for Phil when the patent on the first application was issued. Appreciating the importance of the occasion to Phil, Lippincott took it to the laboratory and with due ceremony presented it to the young inventor. This patent, No. 1,773,980, issued August 26, 1930 (more than three and a half years after application was filed), broadly covered Phil's system of television transmission and reception.

This first issued patent was followed in due course by others covering different phases of the development. Simultaneously applications were being filed for patent protection on all major inventions in all the important countries throughout the world. A stream of foreign patents began flowing in.

As work progressed at the laboratory it became apparent that there were certain key inventions covering important phases of television for which there was no alternative method or device. Then there were patent applications embracing the best, though not the only, way for accomplishing important functions in the system. There were many of these.

Because Phil was pioneering an entirely new field the road was clear for most of his inventions. An unusually high percentage of applications resulted in issued patents with most of the original claims allowed. In a majority of the scores of cases no formidable interferences developed.

However, in two cases of major importance, one covering the "blacker than black" synchronizing pulse, and the other covering the "saw-tooth wave" scanning, a long and difficult struggle ensued. These involved protracted ordeals of testimony-taking and cross-examination on technical matters for Phil, but they were so important to his patent structure that it was necessary to use every means possible to establish his claims. The opposition was equally persistent. The proceedings dragged over many years before the struggle ended and these vitally important cases were conceded to Farnsworth.

(((16)))

Cross-Licensing Arrangement with A.T. & T.

THE CONTRACTS made by the Farnsworth Company with the British and German companies raised the young inventor in the estimation of American companies interested in the electronic art. During this period I was staying in New York and having almost daily contact with the Philadelphia laboratories. One afternoon Don Lippincott came to my hotel and said he had an appointment with George Folk, chief patent counsel for the American Telephone and Telegraph Company. I suggested that he bring up the matter of a possible arrangement between A.T. & T. and the Farnsworth company. I told him that it might be advantageous to invite some of the engineers from the Bell Laboratories to Philadelphia to see the Farnsworth developments.

As a result of Mr. Lippincott's interview I received a call from Dr. O. E. Buckley, president of the Bell Laboratories, suggesting that he and some of the other scientists from their laboratory would like to go to Philadelphia. Arrangements were made for an appointment for the following week, when six of the top-flight engineers of the Bell Laboratories spent the entire day at our laboratories. Among them was Dr. Ives, who had done extensive work in television. I was proud of Phil in his conference with them. As always, he was modest in his presentation, but I could see that they were impressed with the brilliance and originality of his conceptions. Phil was greatly complimented by this attention and hopeful that something substantial would develop from the visit.

Some days after their return from Philadelphia, Dr. Buckley rendered a formal report to Mr. Folk covering their findings, and Lippincott and I were called in to discuss what steps, if any, were to be taken. Mr. Folk was very frank in saying that the Bell Laboratories would like to have an opportunity to inspect the Farnsworth patent portfolio and see if there was anything of value to their company. He was very fair in pointing out that such an inspection carried certain hazards and that we should not present for scrutiny any development which was not fully covered by patent application or formal patent disclosures that would protect the Farnsworth priority.

This was an important decision for us to make. Up to this time we had been careful to prevent disclosures of our portfolio of patent applications to others interested in the art. We were not fearful of pirating, but in this new art we felt that revealing the trend of some of our investigations might lead others into the same paths where nuggets of invention might be picked up before we had reached them in our research.

After careful consideration it was determined that the possible gains outweighed the hazards involved. I asked Phil to send the files up from Philadelphia to New York City.

The portfolio consisted of about a hundred and fifty patents and patent applications. When they arrived one hot afternoon, Lippincott and I bundled the heavy file of documents into a taxicab. Because of their value we did not wish to let them out of our sight. Together we lugged them into the Bell Telephone Building on lower Broadway. After some struggle we got them to Mr. Folk's office.

Mr. Folk received us most cordially. He told us that the patent files would be parceled out by Dr. Buckley to the different departments of the Bell Laboratories for study and report.

"It will take quite a little while to make up a final report for me, probably a matter of four months," he said.

"That will be quite a while for Phil to remain on the anxious seat," I replied. "You know, approval of his work by the Bell Labs has been one of the dreams of his life."

"The disclosures may set some of our boys to thinking," Mr. Folk continued, "and thus Farnsworth may get some new competitors in the fields in which he is working. As I said before, that is one of the risks you run in showing these things to us at this time."

After some further discussion, Lippincott and I left with a sense of real confidence in the fairness of the attitude of this man who represented the interests of the largest scientific laboratory in the world.

Phil felt deeply the importance of our decision. He recognized the enormous impetus that would be given to his prestige, and the Farnsworth company, if Bell Laboratories would place the seal of approval on his work and the American Telephone and Telegraph Company would enter into some sort of license arrangement.

As Mr. Folk had anticipated, it took some four months for Bell Laboratories to analyze the Farnsworth patents and make a formal report. In the meantime Mr. Folk had retired from the office of chief patent counsel for A.T. & T. and W. B. Ballard had taken his place. It would be difficult to find anywhere in the field of business two more fair-minded, kindly men than Mr. Folk and his successor. The vast powers and ramifications of the influence of their great company seemed with them to set a standard of perfection in integrity and fair dealing.

The four months of waiting dragged on slowly for Farnsworth and his associates. Farnsworth had such high regard for the Bell Laboratories that it seemed to him almost too good to be true that they should give formal recognition to the fruits of his years of research and effort. One can readily imagine, therefore, the elation of Farnsworth, and all of us, when word finally came from Mr. Ballard that he had the Farnsworth report in hand. He stated that the Bell Laboratories had found much of interest to them and that he would be glad to discuss the matter with us. This initiated a series of conferences which finally resulted in the preparation of a formal contract between the American Telephone and Telegraph Company and the Farnsworth company giving each the privilege of using the other's patents.

Then came the day when the formal contract was ready for signature. It was agreed by his associates that Farnsworth should be given the honor of signing this document. Phil arrived from Philadelphia early in the morning and came to meet Lippincott and me before going to Mr. Ballard's office. He was in a high state of nervous elation.

When we assembled in A.T. & T.'s offices, on July 22, 1937, Mr. Ballard was delightfully informal and friendly to Farnsworth. He realized that this was really a great moment in Phil's life and made some remarks in recognition of the situation. Phil finally broke down and confessed that he didn't know whether he would have a steady enough hand to sign his name. Ballard, with delightful understanding of the situation, said that he quite understood that this must mean a great deal to

Farnsworth after so many years of work, and in an informal, pleasant way sought to put Phil at ease. Finally the documents were brought in and the moment for signature arrived. Phil fussed around quite a bit, but was able to affix a legible, though squiggly, "Philo T. Farnsworth" to the important document.

((17))

Experimental Broadcasting Station

To KEEP PACE with the rapidly developing new industry it became necessary for the Farnsworth company to make application to the Federal Radio Commission for an experimental television broadcast license. After months of delay the construction permit for the station was granted. A location on a high point, about a mile from the laboratory, was secured. To Mr. McCargar and me it seemed that the station would never be ready for operation. With Phil's characteristic desire to "try out a new idea to make it work better," the station was always just about completed, but "not quite ready." Phil's sense of showmanship, and his desire to have a completely streamlined studio, added to the expense and delay. In the back of his mind Farnsworth had a lurking hope that this experimental station would be a stepping stone to the company's entrance into the commercial broadcasting field. It was a heavy drain on the finances of the company. A simpler setup would probably have served our research purposes better.

This was only a part of the seemingly ever-increasing expense of the laboratory. The monthly payrolls and other expenses skyrocketed to an alarming degree. Circumstances required that

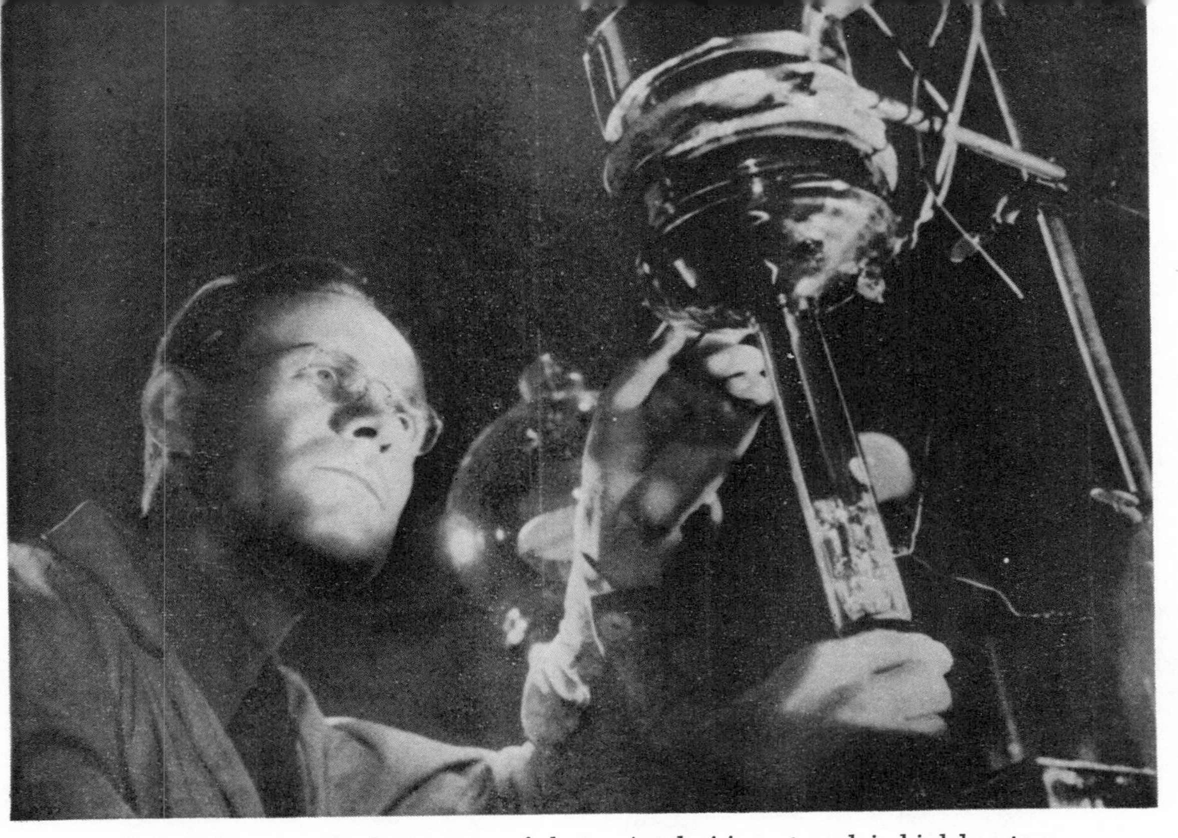

Philo T. Farnsworth, the inventor of electronic television, at work in his laboratory

The first television camera set up in the Green Street laboratory

Farnsworth operating his first portable television transmitter unit

I be in New York to keep in touch with the Philadelphia operations and explore the possibilities of developing some plan for commercialization.

Coincident with the Philco arrangements Mr. McCargar had arranged the purchase of interests of all the original financial sponsors but Mr. Fagan and himself. From that time on the only source of financing was the sale of the corporation's stock. This task fell largely on me and on Miss Helen Baker, who was in charge of the San Francisco office. Miss Baker had long been Mr. McCargar's secretary. When the burden of the Farnsworth development fell on Mr. McCargar and me, she proved to be a tremendously helpful ally. A former Pacific Coast tennis champion, she had a world of friends and influence as well as brains and tact. In helping out with the Farnsworth venture she used the same generalship that made her so difficult to beat on the tennis court.

It was a never-ending struggle to find enough money to meet the expanding demands of the laboratory. Often it was only the heavy load of responsibility we felt toward those who had already put money into our venture that drove Jesse McCargar, Helen Baker, and me doggedly at it to keep the venture afloat. Phil knew that things were difficult, but we had always found the money for him, and he seemed to have an abiding faith that we always would.

The years in the Philadelphia laboratory were years of patient development of the details of the system. It was here that all of the requirements for successful television were brought into sharp relief. Phil was particularly plagued with the inadequate sensitivity of the dissector tube and was painfully aware that the Zworykin iconoscope was superior to his instrument in this regard. Very often on my visits to Philadelphia, Phil and I would have long discussions regarding this deficiency. He recognized that the dissector tube gave a much sharper image than the iconoscope. However, nothing short of perfection seemed to satisfy him. The one advantage of sensitivity was not his, and to him it loomed larger than the comparable advantage of sharpness and clarity that was abundantly in his favor.

In the transmission of motion pictures, where there was plenty of light, the results achieved by the dissector tube were much superior to those of the iconoscope. The contrast was sharper and picture definition was better. However, in the experiments at the studio the light requirements were in excess of what seemed practical for commercial program work. In our discussions Phil usually finished with the statement that neither the dissector tube nor the iconoscope was the ultimate television transmitter tube for studio and outdoor work. He recognized the fact once pointed out by Paul Keston, vice-president of the Columbia Broadcasting System, that to be practical for all purposes the television camera must be able to pick up the last quarter of a football game in the late autumn afternoon on the shady side of the field. This meant that the television camera must be as sensitive as the ones used for motion-picture news shots, if not more so.

Farnsworth sought by various means to correct the deficiencies of the dissector tube. He attacked the problem in several ways, the most promising of which was along the line that was later followed by R.C.A. in the production of their orthicon tube. This development was a pet for a long time and absorbed much of Farnsworth's attention. Though he got some results from it, they were not completely satisfactory. However, the work in this direction did give him patent coverage of the important orthicon tube.

Another line of attack was what he called the image amplifier tube, in which by a most ingenious and subtle process the whole electron image is greatly amplified within the tube before it is scanned at the anode. Many other devices were tried and found unsatisfactory.

Since then, by steady and continuing effort, the sensitivity of the dissector tube and the effective gains in the amplifier have been greatly improved. Also, some of the deficiencies of the iconoscope have been corrected, and a highly sensitive orthicon tube has been developed.

In the Philadelphia laboratories Farnsworth employed from twenty-five to forty technicians and engineers. This was quite a different proposition from the old San Francisco laboratories, where he had at most five or six men working with him. He was loath to give up any executive responsibility, but soon found that the management of so large a group of men was a full-time job. This naturally slowed down the productivity of his own inventions and research.

Mr. McCargar and I both realized the situation but found it difficult to cope with. It took Farnsworth a long time to learn that there are limitations to one man's time and effort, and that there couldn't possibly be enough time available for him to manage so large a staff and do maximum productive work on his own account. It was a serious situation which it took a long time to correct.

As time went on, Farnsworth became much more widely known nationally and internationally. The laboratory seemed to be literally overrun with people from all over the United States, and in fact from all parts of the world, eager to meet the young genius who was regarded as the father of television. It was difficult for Phil to deny interviews to noted scientists from foreign universities, or from American seats of learning, and it was not possible to avoid making appointments with government officials and representatives of foreign governments. It seemed that they always came at the most inopportune time, so that they broke up a carefully prepared schedule that Farnsworth had arranged for himself and his associates.

Some of the situations that developed from the desire of people to see Farnsworth had their comic aspects. I remember one afternoon the chief engineer of the Federal Communications Commission called from Washington to inform us that a representative of the French government was on the train from Washington to Philadelphia and asked if someone would meet him at the specified time at the Broad Street station. Since the Frenchman had met none of our people, and since none of us had ever seen him, the question of identification presented an awkward situation. Frank Somers was sent down to the station to meet him, hoping that he might by some means be able to spot the visiting official. He had no difficulty, because the Frenchman was the only one who got off the train wearing a frock coat and striped trousers.

On another occasion, in San Francisco, a representative of the then great Mitsui Company, which had extensive offices there, called me on the phone. He informed me in clipped English that a learned scientist from the Japanese Imperial University had arrived with letters to him asking that he arrange for the learned doctor to visit with Farnsworth laboratories.

This had not been the first Japanese who had requested introduction to Farnsworth and his work, but none had been introduced with quite so much formality.

The Mitsui representative said he would like to have me and Mr. Farnsworth join the Japanese scientist and some of the company executives at luncheon at the Commercial Club. I told him that Mr. Farnsworth was in Philadelphia but that Mr. Brolly, our chief engineer at the San Francisco laboratories, would probably join me in accepting their invitation.

At luncheon the Japanese scientist, whom we shall call Dr. T., proved to be a toothy myopic fellow with little knowledge of English. He was most agreeable, and with the help of his countrymen the difficulties of language were met fairly well.

He made it plain to us that he was prepared to spend several days at our laboratory if we would grant him the privilege. Since we had little hopes of successfully exploiting our inventions in Japanese markets we could see no harm in letting him do as he wished.

On the second or third day after the visitor had seen the television transmitter and receiver and had spent some time in technical discussion, our engineer inquired of the Japanese regarding the progress of television in his country.

He replied, "We already make your receiver set. When I go home we make your transmitter."

We were not particularly disturbed by this surprising confession, but we were a bit startled by his frankness in stating his intentions.

During his visit Dr. T. formed quite a friendly attachment to Brolly. He finally left after a visit of some four days. When he arrived in the east he visited Farnsworth at the Philadelphia laboratories. He returned home by way of Europe. En route he kept Brolly and me posted as to his journeyings by sending us post-card greetings from spots all the way around the world on his way home. Some months later Brolly got copies of what looked like Japanese scientific journals. The only thing we could read was the name Farnsworth spelled out in English in several places in the pages.

From time to time Farnsworth was called upon to make public appearances before engineering societies and scientific groups. He always regarded such occasions as a waste of time and a tax upon his energies. If possible, he avoided them. He was obsessed with a passion for perfection in public appearances and would never present a scientific paper without weeks of preparation. Although an indifferent speaker, he was able to hold a scientific audience by the brilliance of the subject matter of his presentation.

One of the most difficult of his public appearances was one in 1936 before a joint meeting of the New York Chapter of the Institute of Radio Engineers and the Society of Electrical Engineers. At this gathering he presented a paper and gave demonstrations of his multipactor developments. He took a great deal of pains in preparing this address and spent much time making up slides of charts illustrating his work. He also arranged to present actual demonstrations of his new tubes. A whole stageful of electrical equipment was moved to the Engineers' Club in New York and set up for visual demonstration of the various phases of the operations of the multipactor principle. In fact, the platform looked very much as though Houdini were about to perform.

When Phil and I arrived, about twenty minutes before the appointed hour for the lecture, we were aghast at what we saw. The entire stage was overrun with what seemed to be a good share of the audience. There was hardly standing room around the equipment. Naturally, for a demonstration of this kind there were many wire connections strung about. We could only hope and pray that something was not disconnected or put out of commission by the spectators. Our fears were unwarranted; they were all engineers and had proper regard for the equipment, treading softly and touching nothing.

At the appointed hour of the meeting the hall was packed with the ranking radio and electrical engineers of the New York area. Phil's lecture was more than an hour in length and to me, a layman, it was too abstruse to be interesting. However, it rated as a great success with the engineers. At the conclusion the meeting was thrown open for questions and general discussion. Phil was quick in his responses and generally gave a light humorous touch to his answers. His extemporaneous remarks were much more interesting than his formal lecture. There was some good-natured heckling on the part of a few engineers from competing laboratories, but in answering, Farnsworth came off very much the victor, to the delight of the general audience.

Following the lecture the technical staff from Philadelphia gave a demonstration under Phil's direction of the operation of the equipment on the stage. A high-voltage unit had given the engineers considerable difficulty during the afternoon. There was some doubt in their minds as to whether they would be able to give a successful demonstration of its operation. Bill Eddy (now director of Television Station WBKB in Chicago) had responsibility for this part of the program. As he hooked up the connection and set the tube in operation he was apparently the most surprised person in the room when it functioned without a hitch. His under-the-breath exclamation of "My God, it works!" was clearly audible throughout the room.

This appearance by Farnsworth led to many invitations to address scientific and engineering groups in various parts of the country. Only one, extended by a Chicago engineering society, was accepted. Both the New York and the Chicago lecture received favorable notices in the press on a plane that added much to Farnsworth's reputation as an inventor.

Upon receiving a license for broadcasting from the Federal Communications Commission, Farnsworth and his staff took great interest in the development of television studio equipment and the television transmitter, yet it seemed difficult for them to get a practical demonstration set up. Phil and members of his staff were everlastingly trying out new ideas on the unit being developed at the studio for demonstration purposes.

This situation was peculiarly exasperating to Mr. McCargar and me. We were responsible for financing the venture, and it seemed to us that if we were to continue the operations successfully and bring things to a practical conclusion, we must have some actual proof of the commercial value of television in the form of a smoothly operating broadcasting unit with a picture comparable in clarity to the ordinary movie. In a vague way Farnsworth realized this necessity, but since money for the laboratories had always been forthcoming without any effort on his part, it seemed impossible for him to realize that there could be any real difficulty in keeping up a continuous flow of money into his research program without any tangible commercial results emanating from it.

After months of work and many disappointing efforts on the part of the research staff, and much worry and anxiety on the part of Mr. McCargar and myself, a creditable demonstration unit was brought into operation. This, however, was not achieved without resorting to the drastic expedient of bringing to Philadelphia from San Francisco the two engineers, who by temperament and training were better able to get the maximum practical results from the fruits of our research.

Theoretically Farnsworth recognized the importance of emphasizing the commercial aspect of our development, but at heart he was an experimental scientist and could never resist the temptation to try some new idea before its predecessor had been brought to its full development. It is not to his discredit that this was so. It was an inherent quality in his genius as an inventor. The difficulty we encountered was to stop at a given point in the experimental work, make up a unit representing the maximum achievements to date, and maintain it as an operating demonstration entirely separate from advance developments. It seems a simple and obvious thing to do until one has tried it with a first-class inventor on the premises.

(((18)))

Phil and His Family

NO SMALL PART of Farnsworth's success is due to his charming and beautiful wife, Elma Gardner Farnsworth. She has always been devoted exclusively to their home and the furtherance of his career. There is something in Phil's relations to his home that harks back to the old Mormon tradition of the head of the family being the patriarch. Phil always has felt more than the usual responsibility toward his brothers and sisters and their families. In his early married life he carried this patriarchal relationship into his own family to the extent of dictating his likes and dislikes regarding hats and costume jewelry. Pem bore this with a humorous good will and never took the father-of-the-clan tradition too seriously. She has always been his match in wit and good judgment as far as family matters are concerned.

His parental attitude toward his brothers' and his sisters' families has probably done more harm than good. For a time it engendered a dependence on Phil that was harmful to their own independence of action and initiative. In this case, as in most cases of a similar nature, generosity is likely to have some of its roots in personal egotism. It often does the recipient of favors more harm than good. It is to the credit of Phil's immediate family that they did not impose on this generosity and in due course of time shook themselves loose from its influence.

Phil, too, came to recognize the necessity of independence on the part of his family. They in turn were aware of his great genius and treated him accordingly, even to the extent of spoiling him in many ways. It would probably have been better for him if they had questioned his family authority more.

Phil is instinctively generous and kindhearted, so when his television holdings came to have considerable value, he made special effort to share with those near to him.

Phil and Pem Farnsworth's first child, Philo, Jr., was born in San Francisco in September 1929. In 1931, shortly before they moved to Philadelphia, their second child, Kenneth, was born. Both children were moved to Philadelphia when the laboratories went to the Philco plant. Unaccustomed to the rigors of the eastern climate, Kenneth contracted a streptococcus infection of the throat. Although Philadelphia has some of the best specialists for this kind of ailment in the country, they strove in vain to save the child, and he died after an unsuccessful emergency operation.

Phil was stunned and baffled by this tragedy. I remember being awakened at five o'clock one winter morning in San Francisco by the violent jangling of the telephone, which I answered in a sleepy stupor. Phil's voice at the other end of the line in Philadelphia wakened me as by an electric shock. From the very tone of his voice I knew that something desperate had happened. His first utterance was the anguished words, "Ken is dead." In the confusion I misunderstood him to say, "Pem is dead," and was stunned with the import to his little family. However, as the confused and tumbled words of his tragedy flowed over the wire I realized that it was his boy, rather than his wife, who had died.

This was during the difficult years of the depression, and Phil and his family were living on the modest salary that we were able to provide for him. I realized his need of funds and assured him they would be immediately forthcoming. The young parents wanted their child buried in the home soil of Utah. Pem took the sorrowful journey alone and was met at Provo by her father and other members of her family.

The loss of his younger son made an indelible impression on Farnsworth and led him to study medicine as an avocation and to follow with live interest the new discoveries having to do with the deadly pneumococcus and streptococcus infections. He formed warm friendships with some of the leading physicians and surgeons in Philadelphia. His genius for readily understanding scientific subjects enabled him to acquire a broad knowledge and facile vocabulary of this new subject that often astonished his friends. His knowledge of medicine has led to a certain amount of self-diagnosis that in one with less common sense and good judgment might lead to hypochondria. In later years I have sometimes wondered if this self-diagnosis has not been somewhat harmful in the several sieges of illness that he has endured.

The fact that science could do nothing for his boy when the relentless streptococcus infection closed the little throat left Phil with a sense of the futility of science in its inability to meet this simple emergency of life and death. As a result he spent a great deal of time contemplating ways of adapting his inventions in electronics to medical use. On different occasions Phil discussed the subject at length with members of the staff of the University of Pennsylvania Hospital. One of the doctors, who was devoting his life to medical research, came to New York several times to see me regarding the possibility of raising funds to finance a foundation to endow research work to adapt the Farnsworth inventions to X-ray and therapeutic use. This doctor's particular field was roentgenology, and he felt that such a foundation could contribute much to the welfare of mankind. Many of the Farnsworth inventions may eventually find use in this field. It is a matter of providing the funds to finance patient and painstaking developments.

About two years after the death of Kenneth, Russell Seymour was born. This boy and Philo, Jr., the older son, have been encouraged in original thinking and experimental work by their father and mother. They were roistering, healthy youngsters.

When Phil and Pem first went to Philadelphia, they moved into a two-story brick house of the old Philadelphia style, built close to the next one, which was just like it. It had a deep but narrow and quite pleasant back yard.

When the laboratory was moved from the Philco plant and established in the Chestnut Hill section, Phil took a house near by. It was a large, commodious, and altogether pleasant house located in a fine residential district. It had the advantage of a good lawn and a spacious garden and grass plot in the rear.

It was during this period of the middle 1930's that the trading value of the Farnsworth stock had its biggest rise. The paper value of Phil's holdings went well over the million-dollar mark. This value was largely created by the limited stock available for sale and the constant newspaper and magazine publicity regarding television. That this market value was quite fictitious could easily have been demonstrated if any large amount of the stock had been put up for sale.

The Farnsworths took advantage of this favorable financial situation to expand their mode of living somewhat. Phil bought a Packard car for himself and a coupé of the same make for

Pem. Phil made an effort to take time off for recreation and joined a near-by exclusive country club. He and Pem engaged in a limited number of social engagements.

Pem hired a colored girl to help her with the children and the housework. They also took occasional pleasure trips, the most pretentious of which was a trip to Bermuda. On this jaunt they took the children and a nurse. Altogether it was a pretty expensive jaunt.

Whenever Phil needed money for additional personal expenditures of this kind he usually wired Miss Baker in San Francisco asking her to sell sufficient of his stock to meet his requirements. Since the market was limited, such requests sometimes made it more difficult for us to find the necessary money to meet the laboratory payroll. The most embarrassing instance of this kind was when Phil, on this particular trip, found himself short and telephoned Miss Baker from Bermuda to sell quite a substantial amount of stock, in fact to realize about the same amount that was required to meet the semimonthly payroll.

The market was very soft at this juncture. The laboratory had been under some extra-heavy expenses. As a consequence I was having difficulty in finding enough money to meet the current needs. This added request from Phil for stock sales was almost the last straw. Somehow between us Jesse McCargar, Miss Baker, and I found the necessary money and Miss Baker met Phil's needs. I was inclined to be critical of this thoughtless request on Phil's part but cooled off before he returned from Bermuda. When he got back he was much refreshed, and I thought it best to say nothing about it.

While in the Chestnut Hill house Phil made a determined effort to get some recreation. He set up a badminton court in the back yard and a croquet court in the front yard. Occasionally he would drive up to the Poconos for shooting and snow sports. But he never really went at any of these activities as though he meant it. He just nibbled at recreation; he never got enough for real satisfaction.

He took his meals in much the same way. He was underweight, but try as Pem would to tempt his appetite, Phil usually just picked at his food. He just didn't eat enough. I often dropped in for lunch with the family. In observing his finicky eating habits I often wondered if they were not a manifestation of hypochondria or a sort of misdirected exhibitionism. It used to worry me a great deal, and I often expressed my anxiety to Phil.

I knew that during the long periods of research work in San Francisco and at the Philco plant, and during the first year of his work in the newly established Philadelphia laboratory, Farnsworth had never known what it was to take even a week's time off for a vacation. For years his work was so engrossing that even occasional week ends were no part of his calendar. Such grueling and incessant effort finally showed its effects in more ways than one. I was therefore happy that Phil was now making an effort to take time out for recreation and relaxation.

By an involved chain of circumstances, during the depression years a friend and I had acquired title to an abandoned farm in Fryeburg, Maine. On one of his vacation tours through New England Phil and Pem and the children visited this place and immediately fell in love with it. With characteristic impatience, he burned up the wires telegraphing me in San Francisco regarding means by which he could acquire the property. A modest figure was named, and he took immediate title to the ninety-acre farm, which I had never seen.

Phil seldom does things by halves, and immediately all sorts of schemes were on foot to convert the farm into the Elysium of his dreams. Here he would have a hideaway laboratory. He wanted to raise pheasants and quail, and any other wild fowl that he could propagate to grace the fields.

Each one of us has some pet foible or scheme, usually impractical, but very near to our hearts. Ever since I have known Phil he has had a consuming desire to build an artificial lake. A meandering stream that runs through the Fryeburg property gave opportunity for the fulfillment of this ambition. The brook was planted with trout, and plans were immediately under way for the building of the cherished dam before the frost came. The project was finished by the end of the summer.

I did not see Phil until sometime late in the fall, after the equinoctial storms had brought a flood of rain. When I inquired about his artificial lake, he sheepishly admitted that in his haste he had worked less effectively than the beavers and that the fall freshets had washed his dam away. It was a rather tender subject with him, as he prided himself on doing everything with scientific precision, but in this he had overlooked a few simple laws of hydraulics.

Later, after considerable expense, the dam was rebuilt, and from all reports it is not likely to be washed out again so easily.

Farnsworth established a small but well-equipped laboratory at Fryeburg and spent much of his time there in quiet and uninterrupted research. In the Maine winters he and Pem reverted to the sports that they had enjoyed so much as boy and girl in Utah.

The dam has provided the cherished lake of Phil's dreams, but it was not earthquake-proof. Once when I visited him a heavy shake had put two cracks in the structure and opened a fissure under the foundation which let the water boil out below. Later Phil assured me that it had been repaired without great expense.

As the years went on, Phil, because of his health, came to regard this Maine hideaway as his permanent home.

The laboratory and his study were the center of Phil's interest. The study, of his own design, was a spacious room. A great granite fireplace of native stones of his selection dominated one end. Opposite it was a broad landscape window overlooking a saddle in the blue everlasting Maine mountains. In the center of the room was his desk. All about the room were models of tubes that marked the progress of his research. It was a workroom, yet it breathed an atmosphere of satisfaction in achievement and a restlessness to forge ahead to new discoveries in his chosen field of work. When I visited him there, an open tablet was on the desk. Characteristically on the white page were mathematical equations and squiggled drawings of circuits and tubes.

His laboratory was very simple but adequate to his needs. As in the old San Francisco laboratory, the road between the conception of an idea and its reduction to practical use was a long and hazardous one.

On my last visit three major projects were well under way. Each was as startlingly original and unorthodox as was his first conception of the dissector tube. He was not certain that all or any would turn out successfully.

The facilities of the laboratory were intentionally meager. This was as Phil wanted it. He felt that this simplicity provided him with the atmosphere best suited to original work. He had lost his enthusiasm for a large and highly organized laboratory. That may have been necessary for the refinement of the products of his original invention, but he wanted no part in its management.

The farm and other out-of-door interests gave him relaxation from the intense concentration that his research demanded.

As we walked about the countryside, the youthful enthusiasm that carried him through the arduous years in San Francisco still burned. His eager mind forged ahead to new discoveries in the field of electronics.

(((19)))

Financial Problems—Unexpected Banking Support

THE FINANCING of the Farnsworth research program is of interest here because it contributed to the success of the young inventor's dream.

From the time of the original arrangement with Phil and up to the present, it has never been necessary for him to give more than casual consideration to the funds required for carrying out his extensive development. He has been particularly fortunate in having sponsors who, with a great deal of patience and tenacity, have backed him to the limit with an unswerving faith in the soundness of his idea and a vision of its potentialities.

Of course, when I first told the Farnsworth story to Mr. Fagan, neither Phil nor I had any thought of such a long-drawn-out program of research as actually developed. At that time we were seeking $25,000. (Phil thought $12,000 sufficient.) With that sum we believed it would be possible to develop a television system capable of transmitting acceptable pictures for public entertainment. Ignorance on my part, and Phil's inexperience, accounted for this foolhardy assumption. It took thirteen years and more than $1,000,000 in actual cash before the goal was reached.

During the final years of research the Farnsworth laboratories were spending monthly an amount in excess of the sum that Phil, in May 1926, thought would be necessary for the production of an operable television system.

The $100,000 necessary for research expense previous to the spring of 1930 had been provided by personal contributions from the original trustees and partners. The work had gone on quietly—in fact, almost secretly—which was due to some extent to a sort of sheepishness on the part of Mr. Fagan and some of the other backers in sponsoring such a wildcat scheme.

Shortly after incorporation of the company in 1929, there developed in San Francisco a broad interest in the activities of the laboratory. As a consequence the stock of the company assumed tangible market value.

As stated previously, after Phil's first demonstration to the backers some of the trustees, headed by Mr. Crocker and Mr. Bishop, thought it would be prudent to dispose of the Farnsworth interests, if possible, to one of the large electric companies for a modest sum that would assure them a profit on their original investment.

This thought was at variance with the desire of Mr. McCargar, Mr. Fagan, Farnsworth, and myself. To prevent such action Mr. McCargar secured an option on the holdings of the other trustees. Later the option was taken up and shared with Phil, Mr. Fagan, and me, which effectively put control of the venture in our hands.

Upon the execution of the Philco contract, that company relieved us of the monthly laboratory obligations, with the exception of the patent expense. After Phil and his staff left Philco and established a separate laboratory in Philadelphia, the full burden of the financing for the next six years reverted directly to Mr. McCargar and me. The estimated program required a monthly budget of $1,500 in the beginning, but a research laboratory, like a government bureau, has a habit of growing, and it was not long before operations were on the basis of $5,000 to $10,000 a month, including the expense of the small staff at the San Francisco laboratory and the fees for necessary patent protection.

Because of the highly speculative nature of the investment, it was not possible or advisable to resort to public financing. Neither Mr. McCargar nor I was willing to make a stock-jobbing enterprise out of it. There has probably never been a more alluring opportunity for blue-sky speculative activities.

There was no overhead salary expense except for the necessary accounting and simple bookkeeping. Miss Baker took care of all this. All our funds went into research and patent fees. In all the years of our association with the venture neither Mr. McCargar nor I ever received any salary or fees as compensation from the company. The only reward for our time and effort was the accretion in value to our original holdings in stock. Mr. McCargar, as president of the company, had one inflexible rule that we never violated: we never incurred any obligations, no matter how pressing the need, unless money was on hand in the bank to meet them.

As a California corporation, the company was subject to the very rigid rules imposed by the Commissioner of Corporations of that state. During all the ensuing years of development, and up to the time that the Farnsworth company was expanded into a manufacturing organization, the stock of the company was held in escrow subject to the supervision and control of the state commissioner.

From time to time officials of the company were permitted

to offer small blocks of their individual stock for sale for reimbursement of advances made to the company. We had no large reservoir of capital on which to draw to carry on our efforts; it was necessary to interest other people to join with us in the venture. Since all this was carried on through the depression years, it was not always easy to find anywhere from four to seven thousand dollars on the first and fifteenth of every month to meet payrolls and patent expenses. However, it was done quietly and so effectively that at no time was it necessary for us to curtail the expenses of the laboratory because of lack of funds.

Help often came from the most unexpected sources. The participation of the great banking house of Kuhn, Loeb & Company of New York City was an interesting story. It came about in this way:

When I completed my course of graduate work in social science at Columbia University many years ago, I got a job as executive secretary of the Committee on Criminal Courts in New York City. I was highly flattered when, six months after my employment, I was told that I would be permitted to engage someone to help me. I went about selecting this assistant with the same meticulous care as one would take in selecting a college president, and finally determined upon a newly graduated student from Columbia Law School.

One hot June day I had signed the letter of employment and given it to the office boy for mailing when there breezed into the office an eager youth by the name of Hugh Knowlton, sweaty with zeal and impetuousness, fresh from Yale University and looking for new worlds to conquer. I was so captivated by his engaging personality and brilliance of mind that I decided he was the associate I wanted to work with me. I hustled him out of the office in order to recover the letter of appointment to the Columbia man before it could be mailed, telling Knowlton to return the next day. In due course he was approved by our director and appointed to the job.

This was the beginning of a close friendship that was interrupted by World War I. The chairman of the Committee on Criminal Courts, Bronson Winthrop, a noted lawyer in New York and head of the firm of Winthrop & Stimson, took a great liking to Hugh Knowlton and induced him to study law at Harvard. Upon the completion of his law course Knowlton entered on a brilliant career as an attorney in New York which eventually resulted in his joining the banking house of Kuhn, Loeb & Company as a partner.

After World War I, I went to the Pacific Coast, so over a period of several years we had no communication with each other. On one or two trips to New York I made unsuccessful efforts to see him. At the time of his joining Kuhn, Loeb & Company as a partner I saw a news item about it in a San Francisco paper. Later I learned through friends that he was also acting as a member of the old Committee on Criminal Courts which had originally employed us.

One time while in New York I dropped into the office of the director of the committee and was invited to attend one of the very formal monthly luncheon meetings which were held at the old and aristocratic Down Town Association on Pine Street. I was seated next to the chairman, Nathan Smythe, when Hugh, who was rather late, came in and spotted me. At once he walked to my place at the table and greeted me cordially as "my old boss."

Following the luncheon meeting, we chatted and I arranged to meet him the next day for lunch. In the course of the next day's conversation, catching up on each other's activities during our absence from one another, I told him the story of the

Farnsworth development and my part in it. He became greatly interested. When I made occasional trips to New York during the next two years I usually saw Hugh, and he always inquired for a detailed report on the Farnsworth activities.

One day, after a long luncheon in which Hugh had confined the conversation almost exclusively to the Farnsworth situation, we walked back to Kuhn, Loeb & Company's offices.

As I was about to leave him he said, "George, I believe you are approaching the time when you will be in need of some banking assistance. When the time comes I want to discuss the matter with you."

As a result of this casual conversation I made it a point to have Hugh meet Mr. McCargar the next time we were in New York. It was a friendly and pleasant meeting, with no mention of any kind of business relations. I wanted Hugh to understand a little better the nature of the sponsorship of the venture.

Not long after that Hugh invited me to lunch with him in a private room at the Kuhn, Loeb offices. A great part of the afternoon was spent in discussing the history and progress of the Farnsworth company. The next morning Hugh phoned me, asking if I could meet him at his office. I went down immediately and was taken by Hugh into the private office of one of the senior partners.

Hugh was somewhat excited. As soon as we were seated he said, "George, I have taken up the matter of the Farnsworth company with my partners this morning and we have decided that if you wish we will be glad to put all of the resources of the Kuhn, Loeb organization back of your venture. I believe we can be of real service to you if we can find a fair basis for participating in your company."

This seemed a little better than having one's dreams come

true, because I had vividly in mind the humble beginning and years of patient striving that had led us to the point where one of the great banking houses was voluntarily offering us a helping hand.

Hugh went on to say, "I don't know what would be a fair basis, and I suggest that you get in touch with Mr. McCargar and have him come East to discuss the matter."

I said that I would write him immediately, and Hugh replied, "I suggest that you wire him."

There are times when a stroke of good fortune seems a little too good to be true. As I left the Kuhn, Loeb offices I found it necessary to steady myself in the hallway by leaning against the wall before venturing into the street. I had imagined that at some time Hugh, or his firm, might offer some constructive suggestion, but at no time had I anticipated or hoped that a great conservative banking house would offer direct help or participation by its partners in our venture.

As soon as I returned to my hotel I telephoned Phil the astonishing news. He, of course, was elated, but he had some reservations. "You aren't going to let them get control, are you?" he asked.

I assured him that that was not the idea, but that we had gained some powerful support. "I want you to meet Hugh," I told Phil, and added laughingly, "You know he is one of my finds and protégés, as you are."

Phil was eager to visit the offices of Kuhn, Loeb & Company and meet Hugh, as I had suggested. This was another rung in his ladder to success and he was anxious to step up on it.

Within a week Mr. McCargar came on to New York. He and Hugh and I set about to work out a satisfactory arrangement between the Farnsworth company and Kuhn Loeb &

Company. There was nothing in our discussions that would justify the popular conception of the great banking house swallowing up the little company from the West.

"What we want to do," said Hugh, "is pitch in and be of some help to you. I am sure we can be very useful to you at this juncture. We only want a small participation for the services we render. We have just completed negotiations for a similar group and have made a fine contract with the best company in the field of their interest."

The plan suggested by Hugh was very simple, giving to Kuhn, Loeb & Company a modest compensation for their services and providing that the partners were to have an option on a limited block of the Farnsworth stock at a fair price. This option was later exercised at a time that was most advantageous to us in our program of financing the activities of the company.

While the responsibility of meeting the constantly mounting payroll still remained with McCargar and me, we felt that at last we were headed somewhere in the eventual working out of our plans to some practical conclusion.

From the moment the arrangements were made we were always welcome callers at the Kuhn, Loeb offices. None of the partners was ever too busy to sit down and discuss in great detail the difficult and knotty problems of the company's development. Naturally most of the discussions were with Hugh, although Louis Strauss spent a good deal of time with different members of our organization.

From our experience with this firm I am convinced that sincerity, untiring effort, and a meticulous integrity and honesty of purpose have given it the position of power and influence which it enjoys.

The association with Kuhn, Loeb & Company was interesting and valuable to Phil. There were many enjoyable conferences at 52 William Street between the representatives of great financial power and the young inventor.

Negotiations of the licensing agreement between ourselves and American Telephone and Telegraph Company came up shortly after the arrangement with Kuhn, Loeb had been completed. In this, as in other things, the help of Hugh and his associates was most effective.

(((20)))

Television Network Possibilities

FARNSWORTH, in all his thinking regarding the commercial success of television, considered that network broadcasting was inevitable. He considered that this would come about as a natural sequence once he had succeeded in making television broadcasting and reception practical.

He was familiar with the history of radio and felt that television would follow somewhat the same pattern. When radio broadcasting was new, commercial sponsorship of programs came as a surprise. Originally it was thought that broadcasting stations and programs would have to be supported by the receiving-set manufacturers. However, soon after the first station, KDKA, built by Westinghouse in Pittsburgh, went on the air, commercial advertising automatically solved the problem of paying for the station. It did not take long for shrewd men to sense the great commercial possibilities of radio broadcasting. In a short time far-flung networks of associated stations covered the nation. Programs originating in New York, or other centers, were piped by a telephone wire network to all the stations in a national hookup. A radio broadcast requires seven voice channels to send a program from one station to another. These facilities are provided over the wires of the American Telephone & Telegraph Company on a rental basis.

Such service has been highly profitable to the telephone company. Phil reasoned that such companies as A.T. & T. and R.C.A. were aware of the commercial possibilities of similar services for television networks and would take the necessary steps to provide them.

In television the band of frequencies required in the radio spectrum is much broader than in sound transmission. In fact, the channel required for the successful transmission of a 441-line picture is comparable to that needed for 750 two-way telephone conversations. Naturally, this constitutes a much more difficult problem for station-to-station hookup than that of the radio networks.

Farnsworth visualized the solution of the problem by two methods; one by radio relay, and the other by a wired circuit. Both methods presented difficulties quite beyond the facilities of the Farnsworth laboratory; therefore, while it was an ever-present source of some anxiety to Phil and his backers, they intuitively assumed that either R.C.A. or Bell Laboratories, or both, would provide the solution. It seemed in line with the general trend of their communication developments. This faith was well placed. Bell Laboratories have provided the coaxial cable, and the research laboratories of R.C.A., in co-operation with General Electric, have developed a very simple system of broadcast relays from station to station so that there is now available adequate means for chain television broadcasting either through coaxial connections or through a network of radio relays.

This development was a part of the general research plan of Bell Laboratories to provide simpler carriers for the telephone circuits. It is the result of a long, painstaking effort to put into one single channel the facilities that make up the open-air telephone lines from city to city.

The coaxial cable in its essence is a simple thing. It consists of a flexible copper tube with a fine wire suspended in the center and kept from contact with the tube by insulating washers placed at intervals of a few inches. The terminal facilities are much more complex. It is here that the skill of the Bell engineers came into play to provide this new medium for intercity communication. In addition it was necessary to develop stable and efficient relay links twenty-five miles apart throughout the length of the cable. Also, means had to be found to produce a medium capable of meeting the exacting engineering requirements at a cost that would not be prohibitive.

A good deal of newspaper publicity attended the first experimental coaxial line, which was laid between New York and Philadelphia. Considerable stress was placed on the enormous cost involved in the laying of the cable. Those interested in delaying television seized upon this expense as one of the reasons why television would not be practical for many years to come.

At this time Farnsworth and his associates were really the only people in the radio or electrical industry whose particular interest was directed solely at furthering the immediate commercialization of television. In the middle 1930's sales of radio sets were booming and practically all of the radio manufacturers were making money. The broadcasting companies also were operating profitably and, naturally, did not look with particular favor upon any disturbance of this highly satisfactory situation. Therefore, while the announcement of this cable gave promise of a solution of the chain broadcast problem, it provided the proponents of delay with a tremendous weapon in molding public opinion to the idea that television was "years away." It is true that the expense of the initial cable from New

York to Philadelphia was large, but there was no proof that the development expense involved would be repeated when connections were made with other cities.

Shortly after this first stretch of coaxial line was laid, Bell Laboratories invited Farnsworth and some of his associates to the Philadelphia Bourse Building to view a private showing of the transmission of a television picture from New York to Philadelphia. Several of the most eminent engineers of Bell Laboratories were present, among them Dr. Ives, the great experimenter in television for the Bell System. We were ushered into a small office where the picture was shown. The transmission was of a motion picture of a horse race. The received image was of great clarity and surprising sharpness considering the fact that for this test they were using somewhere between 200- and 300-line detail. I remember particularly the sheen of the flanks of the beautiful animal that won the race as he was led up for the award.

After the demonstration Farnsworth and members of his staff held quite an extended technical discussion with the engineers of the Bell Laboratories. The Bell engineers showed us a duplicate of the cathode-ray tube upon which the picture had been received. They told us that this tube had cost the laboratories $10,000 to construct. It stood fully five feet high and had the most astonishing array of apparatus in the stem. There were all sorts of mechanisms to keep the vacuum constant, to check the performance of every element in the tube, and generally, to make the test transmission foolproof. To one who reads casually in the newspaper the story of a demonstration of this sort, there is no realization of the patience and skill that have gone into the successful accomplishment of the results. In this case the very expensive tube had been used instead of the ordinary $25 or $50 cathode-ray television receiving tube because they wanted to be able to trace exactly the cause of any distortion or interference that might show up in the picture. It was a perfect test of the efficiency of the cable and its relay stations.

In taking such meticulous care in an experimental test of this kind the Bell Laboratories were only following the essential routine that must precede commercial use of a new development in order to make it foolproof and serviceable at all times. In this demonstration we were aware that we were seeing one of the early tests of a city-to-city transmission of television which would undoubtedly be the forerunner of vast networks of similar cables that would cover the country and bring news events instantaneously to vision throughout the land. As to whether this is the ultimate solution of the television network problem remains to be seen. It is possible that other and simpler means may be devised.

It had been proved, first by our early experiments in San Francisco, and later by the National Broadcasting Company, that television programs can be sent for a short distance over an ordinary telephone circuit. Farnsworth and his engineers believed that through the use of the multipactor tube open-wire circuits could be developed to carry television signals long distances by the use of a simple relay system with booster stations at frequent intervals. At the time some limited experimental work along this line was done at the Philadelphia laboratories in co-operation with engineers of the communications division of the Canadian Pacific Railway.

Demonstrations such as that given by the Bell Laboratories were a source of inspiration to Farnsworth. As we drove back to the laboratories from downtown Philadelphia there was a constant flow of talk on the potentialities of this great new development which we had seen. Farnsworth has always had a great deal of respect for a wired connection and has often said

One of the first image dissectors

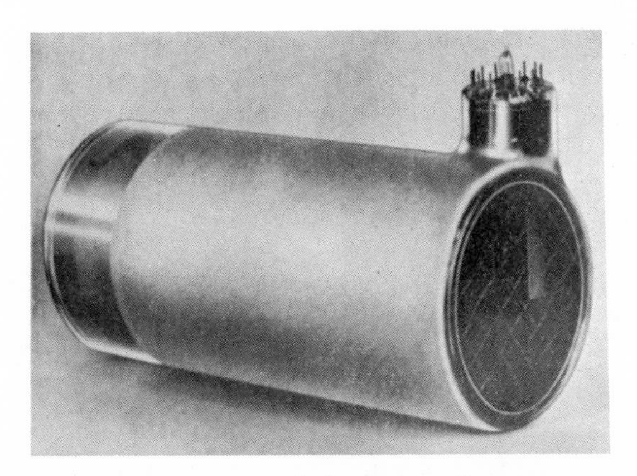

Recent version of the image dissector

First design of the multipactor tube

Recent design of the multipactor
tube

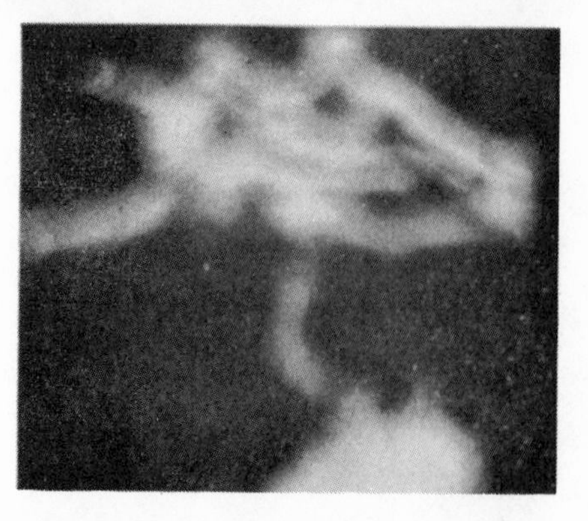

Detail = 50 lines, / 2,500 picture elements

Detail = 100 lines, / 10,000 picture elements

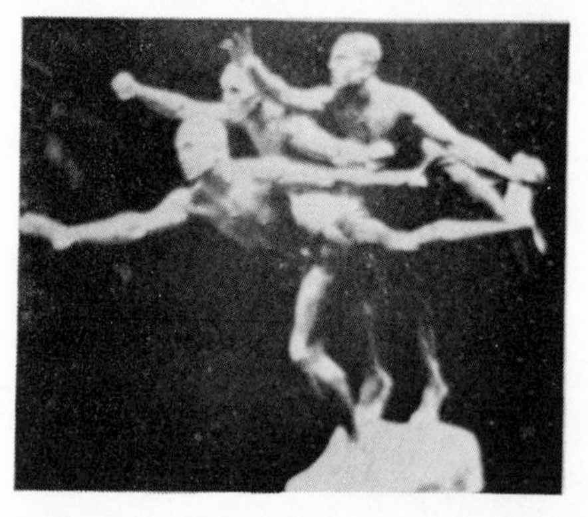

Detail = 200 lines, / 40,000 picture elements

Detail = 400 lines, / 160,000 picture elements

Original photograph

Pictures of photograph through screens of varying mesh, showing resolution obtained

that there is nothing more dependable than a good wire. He has never given up an early conception that he had of piping television to the home either through the facilities of the light and power system of a city, or over telephone wires. At this time he was anxious to try out an experiment of piping a television picture from our laboratories to the nearest telephone exchange and then having it sent from the exchange by ordinary telephone circuit into the subscribers' homes served by that exchange. Certain experimental calculations that he had made indicated that good reception could be had in any home within a two-mile radius of the exchange. This experiment was never carried through because of the press of other work, but Farnsworth still thinks it may be a good idea.

((21))

Standards Proposed for Television Broadcasting

When automobiles were first produced there was no established standard of performance which all manufacturers had to reach before selling their product to the public. As soon as a car was able to chug along a road on its own power it was offered for sale in competition with other automobiles and with horse-drawn vehicles. Sometimes they ran satisfactorily, often not. Competition, engineering advances, and buyer demand for improvements bettered the product from year to year. The result has been the healthy growth of a dominating industry which has brought to us, at a reasonable price, a product of surpassing excellence that has wrought marked changes in the American way of living.

In varying degrees the pattern of the commercial development of the automobile has been followed in the introduction of most new inventions.

The very nature of the art prevented the introduction of television from following the accepted pattern. In considering commercialization it was fundamental that every receiver set sold should be able to tune in on programs broadcast from all transmitters. It became apparent that to accomplish this definite standards of broadcasting must be established by the federal authorities in control of the air waves. It was wisely determined that until such time as suitable standards could be set, television broadcasting should be maintained on a wholly experimental basis. This arrangement gave those engaged in the development of the new art freedom to experiment with any and all types of transmission. It provided testing grounds for proving out what was best, whether it was by the mechanical scanning disk, revolving mirrors, or purely electronic methods.

Another factor in establishing standards was the quality of the picture. Through motion pictures the public had been educated to a high standard of excellence in visual entertainment. It had forgotten how jumpy and how poor in detail the early "flickers" were. Only when we see one of the old-time silent pictures presented as a matter of comic interest do we recognize how great has been the improvement. The public would not now be satisfied with the low quality of picture that was prevalent when the first fabulous fortunes were made in the motion-picture business.

From the first Phil recognized that, to be successful, his television pictures must compare favorably with motion pictures in excellence of reproduction. As mentioned in a previous chapter, in the San Francisco experimental work Phil and his staff made tests of photographing pictures through screens of varying mesh to determine the amount of detail required in a picture to give the desired clarity. As a result of this test he set somewhere between 400- and 500-line detail as the goal to be achieved. The struggle to increase this picture sharpness was long and difficult.

Phil's determination in this regard was his first effort at setting definite standards for television transmission. Later it developed that this problem of picture clarity was but one of the factors entering into those standards. As time went on and the results achieved by the R.C.A. laboratories and others experimenting in television became known, the Federal Communications Commission gave recognition to television experimental work. This was brought into sharp relief as the Commission checked on the operations of the three principal experimental television stations, those operated by Farnsworth, R.C.A., and the Philco Company.

In the years from 1935 to 1937 the Radio Manufacturers Association began to take cognizance of television, and the Federal Communications Commission suggested to the Association that a committee be appointed to recommend standards of television broadcasting for adoption by the Commission. Accepting this suggestion, the R.M.A. appointed a committee headed by Albert F. Murray, chief of the television developments of the Philco Company. The committee was composed entirely of engineers and was at first confined to representation from five companies—Farnsworth, Philco, General Electric, R.C.A., and Hazeltine Corporation. Phil acted as the representative of our laboratories.

At its first meeting the committee established, as a fundamental principle guiding its deliberations, the rule that in all standards they would adopt the device, or circuit, that would give the most satisfactory picture regardless of the ownership or control of the patent covering it. This rule was scrupulously adhered to throughout the committee's deliberations. On some points Farnsworth found himself in agreement with R.C.A., and on others he favored the principles proposed by Philco or Hazeltine.

There were many points involved in setting up standards of

telecasting to meet the rigorous requirements for public acceptance of television as a medium of home entertainment, but there was little difficulty at this time in agreeing that a 441-line picture was adequate [1] and that transmission should be at the rate of 30 pictures per second.

While standards had to do with broadcasting, the size and shape of the synchronizing pulses depended on the design at the receiver. The task of the committee was to select from the developments of all those working in television the simplest, most dependable, and best device for the specific function throughout the whole system. Such questions as polarity of transmission, background information in the picture, single side-band transmission, spacing of sound carrier channel width, and many others had to be determined. All took careful study, fairness of judgment, and practical and experienced engineering knowledge to decide what was best.

The standards were not arrived at hastily. Deliberations were carried on in regular meetings held at intervals over a period of more than four years. In addition to the formal meetings there was much in the way of informal discussions and papers in engineering publications. Whenever a meeting was called, Phil always made it a point to call a meeting of our own engineers and go over with them the technical aspects of the problems coming up for discussion at the R.M.A. meeting.

As the work of the committee progressed, it became apparent that the art had arrived at a point where it could be stabilized and major standards crystallized without harm to future developments and growth.

Phil's dream of electronic television superseding the mechanical methods had been abundantly realized. The committee found no place in their discussions for consideration of

[1] Later this was increased to 525 lines.

equipment involving mechanical moving parts. From meeting to meeting different phases of the art were scrutinized closely. As the months passed, sound conclusions emerged governing the whole field of visual broadcasting and reception. It is doubtful if the technique of a new art was ever arrived at with more careful deliberation.

In later chapters we shall give further consideration to this committee's work.

(((22)))

Program Experiments

FOLLOWING Dr. Zworykin's visit to the Farnsworth laboratories in 1931, we were honored by a visit from David Sarnoff, president of R.C.A. At the time, Phil and Mr. McCargar were in Philadelphia negotiating the Philco contract. Sarnoff seemed impressed with what he saw at our laboratories but told me that he felt Dr. Zworykin's work on the receiver made it possible for R.C.A. to avoid the Farnsworth patents, and that at the transmitter they were using a mechanical mirror device that he thought would equal the results which we could obtain by our dissector-tube camera. He said that their laboratories at Camden were being reorganized and that he was putting his influence behind the developments of Dr. Zworykin. After this visit we felt more certain than ever that the R.C.A. laboratories would give us stiff competition.

Naturally, we were not permitted to know the details of the R.C.A. developments, but from time to time announcements were made to indicate that they were making splendid progress. In due course the Zworykin iconoscope, as the heart of the R.C.A. television camera, was announced. It is a most ingenious tube based on a principle somewhat different from the Farnsworth dissector tube. In construction it is much more complicated than the dissector. Theoretically it has much greater sensitivity than the Farnsworth tube because the electrons from the picture elements are released a point at a time, rather than by a continuous flow from all parts of the entire image as is the case in the dissector tube.

In the Zworykin tube the cesium surface of the cathode plate is deposited in island units, each distinct and separate from the other. During the period of a one-time scanning of an image, that is, 1/30 of a second, the electrons emitted from each element of the picture are stored up on the insular areas and released as the next scanning beam passes over it. In the dissector tube there is no storage during the scanning interval; electrons are flowing constantly from each element in the image. It is this storage process that gives the iconoscope the advantage in sensitivity over the Farnsworth device. However, the insular areas had a tendency to store up excessively large charges of electrons on the brighter part of the picture, which leaked over to the adjacent islands on the darker portions; this evened up the light values and destroyed sharp contrasts, and as a result, fuzzed the detail of the image.

In the Farnsworth dissector tube there was no such spilling over from the light to the dark; therefore a picture produced by the dissector tube was of better definition and sharper contrast than that produced by the iconoscope. It became apparent that the Zworykin iconoscope was a superior instrument for use in studio and outdoor pickup where lighting was a problem. In motion-picture work and bright sunlight, the dissector tube had an advantage.

While little was said openly about the rivalry of the two laboratories, it was obvious to everyone that there was a definite race between Farnsworth and Zworykin in their television developments.

During a period when the Farnsworth laboratories were quartered at the Philco plant, Farnsworth and his engineers quite often picked up the R.C.A. image from the experimental laboratories at Camden. It is remarkable what an engineer can find out by studying such an image. Farnsworth discovered many things that encouraged him. He also learned things about the Zworykin picture that were disquieting. He and his associates were very complimentary about the work of Dr. Zworykin and freely expressed their admiration of his results.

Phil seemed to enjoy this rivalry. It stimulated his inventive efforts. When he found something in the Zworykin picture that showed his own to a disadvantage, he was always hot on the trail of something to surpass it. I believe that Phil's ideas on what later became the orthicon development, and his image amplifier dissector tube, were the results of his determination to outdo what he saw in the Zworykin picture.

During the daily work on improving the television image it became apparent that ordinary line-after-line scanning of the image caused a definite flicker which would have to be removed if television was to meet motion-picture standards in pleasing the eye. To correct this the idea of interlaced scanning was conceived. Farnsworth worked out two methods of accomplishing it. However, as the patent applications on both methods followed their course through the Patent Office, it was disclosed that priority on the preferable one had been awarded Dr. Zworykin. It is to Farnsworth's credit that in the Committee on Standards he voted for the adoption of the Zworykin method of interlaced scanning over his own alternate scheme because he thought it would operate more simply and effectively. And this in spite of the fact that it gave Zworykin a patent controlling an essential element of television.

Interlaced scanning is one of the most subtle and precise operations in the whole art of electronic television. It works both at the transmitter and at the receiver. The whole complicated and accurate operation must be accomplished within an extremely limited area. The odd number of lines in picture scanning was adopted to meet the requirements of interlacing. Transmitting at the rate of 30 pictures per second by the interlaced method means that each image has to be gone over twice in the lateral scanning, the first scanning laying down the odd lines, 1, 3, 5, etc., and the second scanning laying down the even lines, 2, 4, 6, etc. Those of the second operation must be laid down interlaced exactly between the lines of the first scanning. If the picture is broken down into 441-lines, this means not only that it must be scanned at the frequency of 30 per second and the lines picked up and laid down in absolute parallel order, but that they must be interlaced with minute accuracy so that the lines of the second scanning of each individual picture fall exactly between the lines of the first scanning.

When this interlace was first tried out in our laboratories I could only marvel at the nicety with which it was done. I was shown a picture of 221 lines. A knob on the control panel was turned and the interlacing of 220 additional lines entered the field. The result was a smooth, uniform field without jumpiness or flicker.

In the neck-and-neck television race, interlace was a score for Zworykin. To counteract that, Farnsworth had the satisfaction of knowing that his straight-line scanning, his saw-tooth wave form, and the blacker-than-black synchronizing pulse dominated the use of the iconoscope and scanning at the receiver.

While the research laboratories of R.C.A. and Farnsworth were definitely rivals, there were certain elements where there could be no rivalry. We were not equipped technically or financially to compete with R.C.A.'s National Broadcasting Company, or any other company, in the field of broadcast technique, programs, and entertainment.

For a time shortly after our laboratories were established independently in Philadelphia, Farnsworth had ambitious ideas regarding the development of an experimental broadcasting station as a center for our experimental program work. Both Mr. McCargar and I felt that this was quite beyond our pocketbooks and very much outside the scope of the Farnsworth program. We were convinced that we were not fitted for the entertainment field and that when actual commercial television arrived people experienced in radio programs and motion-picture production would have at their fingertips the technique and facilities that would take us months, if not years, to acquire.

However, in the studio at our station in Philadelphia, Phil did succeed in doing quite a bit in the way of experimental programs. He employed Bill Eddy, who later gained considerable prominence in the development of television programs at the National Broadcasting Company and other studios, to direct this work. Eddy made some interesting experiments in lighting for television and in the operation of two cameras picking up studio productions. He also made some tests of fake background with miniature sets. Such tests did little more than prove to our own satisfaction that our television camera was a practical device for the transmission of studio programs.

At this time we had two cameras in operation, in addition to the telecine channel for the transmission of motion pictures. Monitors were so arranged on the control panels that the chief operator at the panels could see what sort of picture each of the cameras was picking up and alternately pipe each into the main channel for actual transmission to the receiver. The man at the monitor was the "mixer" and exercised virtual control over the program going out on the air. Technique was developed for smooth fading from the closeup to the long shots and the reverse. Eddy worked out a very clever miniature revolving globe with the appropriate announcement of "The Farnsworth System Presents." He had quite a genius for showmanship and originality. It was unfortunate that we were not in position to take full advantage of it.

A great deal of attention was given to lighting effects to cut down the light requirements. By the proper use of back lighting and other devices to accentuate the contrasts in features, we were able to increase the effective sensitivity of the dissector tube manyfold to the point where the actual light requirements for transmission of studio programs was something less than that necessary for the taking of colored motion pictures.

Here, I believe, for the first time serious attention was given to make-up for television. Later all of the laboratories gave a great deal of publicity to this important problem and many fantastic news stories of make-up requirements were published in newspapers and magazines. The Max Factor Company, prominent Hollywood make-up artists, became interested and on several occasions had a representative at our Philadelphia laboratories to assist in our experiments.

Because the cesium surface on both the dissector and the iconoscope tube is peculiarly sensitive to the infrared portion of the spectrum, red televises white, quite in contrast to what happens in photography, where red photographs black. A mixture of blue in the make-up provided the proper effects where shading was necessary to deepen the facial contrast.

One amusing incident occurred to illustrate the fact that red televises white. A prominent newsreel service came to Philadelphia to photograph the operation of the television studio and, if possible, to take a motion picture of an actual television picture at the receiver. For the purpose of news interest a couple of boxers were to put on a sparring match on the lawn by the studio building. One of the boxers had on brilliant red trunks. I was watching the television receiver when the first test was made and was very much startled by what looked like a contest between a representative of a nudist colony and the boxer in the black trunks. In the interest of censorship it was necessary for the boxer in the red trunks to change to some other color before we could make a motion-picture record of the received image.

It was also found that certain fabrics were transparent in television transmission. This discovery was made when we were about to put on a demonstration for newspapermen. We had engaged the services of a toe dancer. She brought two costumes with her, one of them of a light material such as toe dancers often wear, which she put on for rehearsal before the newsmen arrived. When the picture came through on the receiving set, instead of presenting a girl dancing in a fluffy skirted costume, we had a dancer performing in virtually nothing. Fortunately she had on tights. She was told that we thought the other costume would be more appropriate. She changed, and the show went on.

In many of our tests we found that a certain type of red hair televised exceptionally well. I happened to mention this one time to a reporter from one of the major news services who was witnessing a demonstration. He wrote up a catchy article about it which was widely syndicated throughout the country. I was quoted as the authority on the telegenic qualities of redheads. For several weeks afterwards I received letters from red-headed girls in different parts of the country aspiring to become television performers. One sent along a lock of her tresses to prove that she qualified.

At this time we also carried on at the studio quite extensive experimental work in the pickup of outdoor scenes. Under ordinary sunlight conditions the dissector tube operated satisfactorily, but there were definite deficiencies when we attempted to transmit pictures from out-of-doors on afternoons of gray days. On the whole, however, we were greatly surprised at the effectiveness of the television camera in the transmission of ordinary out-of-door activities. We rigged up an improvised badminton court on the lawn of the studio and did very satisfactory work in transmitting the game.

One thing that was always a source of surprise and satisfaction was the beautiful way in which television transmitted action. Motion seemed to improve the quality of pictures. The only explanation we could give for this was the fact that the human eye is one of the greatest creators of illusion on earth. In action pictures the eye supplies from memory so much that is not there. This is true of motion pictures, and it was proved to be so to an even greater degree in television pictures. It was remarkable to see the clarity of images of automobiles on the highway half a block away, or the activities in back yards in the neighborhood at some distance.

A frequently mentioned theoretical shortcoming of television was the size of the received image. In conversation with almost anyone who had not seen a television image one of the first questions usually asked was, "How large is the picture?" When the answer was given that the dimensions of pictures were from five by seven to nine by eleven inches, there was usually an expression of disappointment that the pictures were so small. However, our tests in the Philadelphia studio proved to us quite conclusively that pictures of the dimensions named would be adequate for commercial home receiving sets.

This was very well expressed by a friend of mine who handled two of the most popular radio programs on the air. At his request I took him and his family from New York to Philadelphia for a demonstration. The boys at the lab had built up an interesting program to show us. On our way back I asked my friend what he thought of the picture.

He said, "I feel it is definitely commercial. I was agreeably surprised." He added, "When you took me into the room and sat me down in front of the television set I looked at the small screen on the end of your receiving tube and made the mental notation, 'I wonder if these people think they can give me anything of interest on so small a screen.' However, when the picture came on the size seemed to be adequate. This was particularly true of the outdoor shots." That was the general reaction received from most people who came to see the picture.

On another occasion, when we were making a demonstration for the head of the Communications Division of the Canadian Pacific Railway and his family, the same question came up for discussion. Mr. Neil made this comment: "To me a television receiving set is a personal, intimate thing. When I get my set I will want to sit near it and glance up occasionally to see what is coming on the screen. If it is of sufficient interest I'll take time out from reading, or whatever else I am doing, to watch it closely. To me the size of the picture is adequate for the home."

Much as Phil wanted to have our laboratories conduct experimental work on programs, he turned little of his genius to the task of perfecting the details that lay between us and the actual transmission of acceptable commercial television pictures. He was impatient with the shortcomings of the dissector tube and the equipment in general to meet the high standard of picture detail and sensitivity that he had set for himself. As is typical of inventive genius, he wanted to work out the detailed refinements by some new spectacular approach, rather than through the more tedious method of routine engineering. He seemed unable to recognize fully the merits of what he had achieved and to be unwilling to apply the skillful and patient engineering necessary to get the maximum practical results from his inventions. This attitude on his part permeated the Philadelphia organization. It seemed to Mr. McCargar and me to be utterly impossible to keep experimental hands off the equipment that was specifically set up for the purpose of producing the maximum performance that could be achieved by our scientific developments to date.

Two engineers, Bart Molinari and George Sleeper, working independently in the San Francisco laboratories, were fortunately far enough away from this influence to have proper appreciation of the equipment developed up to that time. They were particularly enthusiastic about the dissector tube and had the patience to do the monotonous engineering required to achieve the best possible performance. Working alone, they developed a television camera of beautiful and efficient design around the dissector tube. They got most of the bugs out of the auxiliary equipment and stabilized the circuits to insure dependable operation.

Two of the harassing difficulties met at this juncture were a so-called "S" distortion and the pincushion effect in the picture. These two irregularities were primarily due to the electromagnetic fields of the deflecting or scanning coils. The "S" distortion put a wave in the scanning line so that there was a sort of hump in the middle of the picture. The pincushion effect gave the impression that the picture was stretched at each corner.

By painstaking work Molinari and Sleeper found out what caused most of the distortion and reduced it to a point where it was not objectionable. They also had greater faith in the effectiveness of the dissector tube as an instrument for the camera than did Farnsworth and his associates in Philadelphia. They felt that if maximum results could be obtained in the circuits, the intense lighting requirements, due to lack of sensitivity of the cesium surface, could be greatly reduced, and they set about to prove it.

After many and prolonged attempts to get a satisfactory demonstration in the Philadelphia laboratories, in desperation Mr. McCargar sent both Sleeper and Molinari to Philadelphia to lend a hand in producing a presentable picture. Such a drastic measure, naturally, caused some embarrassment to the San Francisco boys, and resulted in some friction. Engineers generally are about as temperamental and fractious to handle as prima donnas. Anyone who has had much experience with them will recognize that this was an explosive situation and might have resulted disastrously. However, by the use of considerable patience and forbearance on the part of everyone concerned, the advantages that had been gained in the San Francisco laboratory were incorporated in the Philadelphia demonstration setup. As the result of adding the work of the boys from the Coast to the latest developments in Philadelphia, we were able to give a very satisfactory picture.

Phil seemed to be unable to reconcile his scientific aspirations and his desire to solve every problem by invention, rather than by painstaking engineering, with the practical necessities of the situation. The fault was as much that of his backers as his because we had kept him completely sheltered from the hard financial requirements of the situation. We had felt that it was our obligation to provide the money and his responsibility to carry on the research and development program. Whenever there was vital need for a special project the money was always forthcoming, to the extent that there often prevailed in the laboratory something of the same atmosphere that pervades the laboratories of great research foundations and universities, where there is no necessity to turn the research to commercial advantage. It fell to my lot to bear the brunt of the impatience of the backers, who were seeking commercial results, and to do as much as possible to set Farnsworth's program straight with the responsibilities that he owed to his sponsors and those who had put money into the venture.

At times Farnsworth was inclined to be abrupt and impatient when such matters were brought to his attention, and occasionally he was difficult to deal with. I remember one morning in Philadelphia when I pointed out to him the necessity of putting the studio demonstration in presentable shape.

"Phil," I said, "You and Tobie have been puttering with that camera over in the studio for months now. Every time I drop in here, which is at least twice a week, Tobie assures me that it will be together and operating the next time I come over, but it never is! I was just over there and he has it all torn down doing something with it."

"Yes," said Phil, "yesterday I gave him the design of a new circuit to put in. It will help the pictures a lot."

"I know all that, Phil, but that's what's been happening over there for the last three months. Three months ago you and Tobie promised me you would have that camera operating within a week. Now here's what I want you to do, and Jess insists on it. You take Tobie out of that studio and you stay out of there yourself and then assign Frank Somers to get the thing operating."

Phil straightened up, got up from his chair, and said, "This is my laboratory. Neither you nor anyone else is coming in here giving orders as to what is to be done."

"Your laboratory or not, Phil," I replied, "this is one time when you are going to do as I say. If we are to continue to get money to run this place, some attention must be given to getting a commercial picture. You and Tobie won't do it, so Frank will have to." Then after a pause, "I'll sit here and read the paper until you do as I ask."

Phil gathered up some papers and swung out of the room, slamming the door behind him. I had had similar tiffs with him before, and though I didn't know what might happen, I sat down to read the paper and await the outcome.

I remained there about twenty minutes. At the end of that time Phil came in smiling, as though nothing had occurred, and said he had arranged for the necessary changes, and then went on to the discussion of some technical problem that was uppermost in his mind at the moment.

This assignment of reconciling the purposes and will of an inventor to those of a banker, with the latter's practical and realistic outlook on our development, was not an easy one. It was often a subject of discussion between Lippincott and me. On one occasion Lippincott made this observation: "Most inventors never make any money out of the product of their genius for one of two reasons: they either get in the hands of shyster promoters who ruin any possibility of successful commercial development, or, if they are fortunate enough to have responsible hard-headed financial backers, at some critical period they get at loggerheads with their sponsors. This often results in the backers throwing up the whole proposition in disgust. In this case, while Phil has often been unreasonable, you have always been able to bring them into line."

(((23)))

Facsimile and Fog Penetration

As POINTED out in a preceding chapter, before the laboratory moved to Philadelphia most of the original financing group, with the exception of Mr. McCargar, Mr. Fagan, and me, had released their interest in the undertaking. Later their position was taken by others, and both Mr. McCargar and I felt a keen responsibility as trustees of the funds that had been put in.

Money was attracted to the venture by the two outstanding elements in the situation—the great genius of Farnsworth, and the recognized reputation of Mr. McCargar in the banking world. I am convinced that Mr. McCargar's patience was often tried to the breaking point, but he never lost his faith in the soundness of the Farnsworth inventions, or his sense of responsibility toward those who had put money in the speculation. Added to this was a staunch determination to finish successfully anything that he started.

The real test of the strength of the combination that had been built up came after the original glamour of the new development had worn off and things settled down to a humdrum day-in and day-out routine of perfecting the picture and overcoming the obstacles to practical commercialization. On

Phil's part it was a severe strain on his inventive impatience, and on the part of Mr. McCargar and the rest of us it meant a continuance of faith through long years of more and more expensive development. It seemed that the monthly financial requirements were always on the increase, while tangible evidence of progress appeared in a diminishing ratio.

From time to time a financial sponsor would become impatient and sell his interest. Newspaper stories and an occasional financial item regarding the Farnsworth patent structure, or a victory in an interference suit with R.C.A., piqued the interest of San Francisco brokers and speculative investors. This interest was cumulative. As the result of two or three good pieces of news that found their way into the financial columns in the summer of 1935, there developed a lively interest in the over-the-counter market on Farnsworth shares. Because of a limited supply of stock to be traded, a most amazing rise in price developed. The paper profits of all of us were startling. However, neither Mr. McCargar, Farnsworth, nor I cared to take advantage of the situation. We were convinced that only the commercialization of television would bring us any real substantial profits. That seemed some distance away.

At the time this took place I was flat on my back in bed as the result of an injury. I was so harassed by phone calls at the house that it finally became necessary to have the telephone company put in an extension by my bed. One of my good friends, Ernest Geary, came out to visit me and as a joke brought me half a dozen lead pencils and a couple of scratch pads to enable me to figure my paper profits from day to day.

We had no illusions about the difficult developments and industrial hazards that were ahead and preferred to carry along in the usual way of providing the monthly payrolls that were necessary to keep the laboratories going, rather than indulge in any dreams of great wealth. Nor did we think it was necessary for us to engage in wholesale expansion of our laboratory expenditures.

The stock flurry finally subsided and the enterprise, so far as Mr. McCargar and I were concerned, settled back into waiting for the crumbs of development progress that fell from the breadboard of experiment. Month after month the patient plodding stretched through the depression years. The general economic conditions did not add to the ease of our financial problem, or help in providing the monthly payroll.

While this was going on, the R.C.A. laboratories were making substantial progress with their television. Vast resources in equipment and funds, together with their tremendous prestige, made them formidable rivals. We wisely did not venture to enter into competition with them in any field other than that of patent strength. It is possible that R.C.A. has spent easily ten times the amount of money on television that was put into the Farnsworth development. They approached the problem from the broad commercial angles represented in their industrial structure, which included not only the manufacture of receiving sets and transmitters, but also the allied fields of broadcasting and communications. Aside from the limited operations of our experimental station in Philadelphia, and the simple program experimental work carried on for a time in our studio, we made no attempt to perfect the technique of television programs for commercial use.

Newspaper services developed quite an interest in the adaptation of the principles of television to telephotography and facsimile reproduction. Representatives of different news agencies visited our laboratories in Philadelphia and San Francisco to explore the possibilities. The interest became so insistent that I asked one of our engineers to make a blueprint outline of a telephotography and facsimile system for use over wires, employing the Farnsworth dissector tube and other principles of our television system for the purpose. In diagram form the system seemed quite practical, but Phil took slight interest in it. In this instance, as in many others, he showed good shrewd sense. It was a highly competitive field in which there were several alternative methods for the purpose. In fact, as I looked into the problem further, it seemed that every tenth person I met for serious discussion brought up a new system of facsimile.

Another adaptation of the Farnsworth dissector tube which created widespread interest was the possibility of its use as a fog-penetrating device. It had long been known that infrared rays were able to penetrate fog to some extent. Since the Farnsworth dissector tube was peculiarly sensitive to infrared or heat radiation, it seemed that a camera device might be perfected that would register these rays through the fog and thus become a great aid to aerial and water navigation. For some months the Sperry Gyroscope Company co-operated with the Farnsworth company in experimental work along this line. While there has been some newspaper publicity regarding the prospect of successful fog-penetration devices, our experiments seemed to indicate that though the infrared rays could readily penetrate a foglike haze, or fogs where the moisture particles were of minute proportion, in an intense wet fog there was little chance of their getting through. Whether a successful fog-penetration device will be developed still remains to be seen.

There were many other temptations to spend money and effort in the development of what seemed to be profitable and productive side lines of research to adapt the television tube and principle, as well as the multipactor tubes, to other fields. Enticing as some of the projects were, our limited financial resources and the inability of Farnsworth to spread himself out any thinner prevented our venturing into these alluring pastures. Phil used good judgment in recognizing the limitations of his own time and our financial resources and did not wander off from the main objective, as so often happens with men endowed with inventive genius. Many such avenues were properly open to large companies because of their wide and varied commercial interests. For us it was necessary to stick to the main objective, television.

With the establishment of our experimental broadcasting station in Philadelphia, the executives of the Columbia Broadcasting System manifested a friendly interest in our developments, and Phil and I were proud to arrange a special demonstration one Sunday for members of the C.B.S. staff. At this time Columbia had recently employed an Austrian engineer, Dr. Peter Goldmark, to organize and set in motion Columbia's effort to enter the television field.

Dr. Goldmark and Phil developed a high regard for each other's professional attainments and were most congenial in their discussion of television problems. Dr. Goldmark was a tireless student with great abilities as an engineer. He had a fine appreciation of Farnsworth's originality and a gift for the development of simple and practical devices.

As time went on C.B.S. became quite a regular purchaser of Farnsworth dissector tubes for Dr. Goldmark's department. His requirements were very exacting. This had a wholesome effect on the tube department of the Farnsworth laboratories.

Farnsworth recognized that here, for the first time, was an opportunity for his dissector tube to be put to the test by a commercially operated broadcasting company that was fully aware of all the requirements of the day-in and day-out schedule on the air.

Some time later Mr. McCargar and I were privileged to see

a demonstration of the transmission of motion pictures by the Farnsworth dissector tube at the Columbia studios. It appeared to us by far the best television we had yet seen. We were delighted with what Dr. Goldmark and his staff had done in careful engineering to secure the best results from the dissector tube.

(((24)))

Defense of the Patent Department

FROM THE OUTSET both Mr. McCargar and I had believed that it would be bad business judgment ever to attempt to go into the manufacture of receiving sets or into the broadcasting business. Neither of us, nor Farnsworth either, had had any experience in manufacturing or in the distribution of radios, a highly competitive business. It also goes without saying that none of us had knowledge of commercial broadcasting. We were of the opinion that our company would achieve its greatest success by simply sticking to the research and development program, deriving our revenues almost exclusively from royalties on patents. Phil was not entirely in sympathy with this program and often expressed a desire to get into manufacturing if it were possible. He had what he thought were sound and original ideas regarding a manufacturing program, but in this matter he deferred to our judgment and confined himself to wishing.

When the arrangement with Kuhn, Loeb & Company became effective, Hugh Knowlton was at first inclined to share our ideas regarding manufacturing. However, as time went on it became more and more apparent to all of us that in order to secure the maximum return from our licensing program we must have a proving-out field for the use of our patents in a production plant. It would be tedious to recount the months and months of effort that were put in by all of us to find a satisfactory solution to this problem. There were discussions with several of the foremost companies in the radio, electrical, and motion-picture field concerning licenses for the exploitation of television. Other companies and individuals explored with us proposals for financing and launching the company in various commercial fields. Some were alluring, others seemed sound, but we could not come to a wholly practical plan for any of them.

It was a source of amazement and great satisfaction to me to observe the patience and abilities exerted by Hugh and the associates of Kuhn, Loeb & Company to find a solution. At no time was there any talk of a grandiose structure that might come from the imagination of a scenario writer for a popular motion picture. We were faced with the job of conserving and exploiting a million-dollar investment in research work and any further inventive products that might come from the fertile mind of Farnsworth. The cross-licensing agreement with American Telephone and Telegraph Company had given us a strong trading position. The publicizing of Farnsworth's efforts in this country and abroad had made the Farnsworth name a valuable asset.

The final result of months of study and discussion concerning the future course of the company was the decision to employ an outstanding leader in the radio manufacturing and distribution business to head up a radio and television manufacturing company. This company would be prepared to put on the market a complete line of television and radio receiver sets, as well as to be in position to supply transmitters for broadcasting stations. This determination was made late in the summer of 1938. In order to carry out the program it was necessary to acquire manufacturing facilities and float a stock issue of sufficient proportions to buy the plants and provide adequate working capital.

While Phil was very much at sea in the discussion of the plans, he was at heart quite happy that his original conception of building a great manufacturing company around his patent structure was likely to come true. However, as the plans took more tangible form he was a bit frightened by their magnitude and the responsibilities they involved. He was in turn elated or unhappily confused by what was going on. This attitude was accentuated by the fact that at this time his years of highly concentrated work had begun to tell on him nervously to the extent that we were fearful he might suffer a nervous breakdown, so while the plans were carried on with his full knowledge of what was going forward, he took but a minor part in their development.

Phil was at this time engrossed with some very baffling tests on his image-amplifier tube which were causing him much concern because he couldn't seem to make electrons jump through the ropes the way he wanted them to, and after the pattern which his mathematics indicated they would.

In the midst of Phil's confusion with this problem, and our proposed change of corporate setup, there came from Washington a request for him to testify before the O'Mahoney Temporary National Economic Committee, commonly known as the Monopoly Committee, which was inquiring into the possibilities of patent monopolies.

A part of this inquiry revolved around the operations of the Patent Department. There were rumors that plans were under way for a complete overhauling of this department and reorganization of the patent procedure. The Commissioner of Patents and his associates, naturally, were disturbed. Since Phil was an outstanding example of a poor young inventor achieving recognition in a field where powerful interests were at work, his testimony would be valuable. Phil's treatment by the department had always been so fair that he was eager to appear in its defense.

I went with him to Washington for the hearing. It was held in the Caucus Room of the Senate Building. This was a large chamber elaborately fitted up with a big thronelike table for the members of the committee, each member being provided with a microphone. At right angles to the committee dais and leading down from it to the center of the room were long tables for members of the press, waiting witnesses and their counsel, and privileged spectators. A table at the right was given over to the counsel for the appearing witness and was fitted up with microphone and room for secretaries. Back of the throne were facilities for the use of the committee's staff. This elaborate array occupied one half of the chamber, the other half being reserved for the public. Someone remarked to me that this room was the political sounding board of the nation. The arrangements and the fittings lived up to the definition.

We appeared at the hearing as Dr. Jewett, of the Bell Laboratories, was testifying before the committee regarding the patent setup of the great Bell System. He explained fully the activities of the Bell Laboratories over which he presided and their relationship with the parent company, the American Telephone and Telegraph Company. In the course of his testimony there developed one of those strange instances where a simple statement by an honest scientist was distorted to bedevil two great corporations.

Dr. Jewett said that the expense of $10,000,000 a year for

the laboratories was justified because the new devices developed by them effected huge savings to the Bell System and added much to the efficiency and dependability of its service. To illustrate the point he said that the laboratories had recently developed a very efficient tube for use in their relay stations; that previously the best tube they had had for this purpose had an effective life of 1,000 hours of service, whereas the new tube gave more than 50,000 hours of use. He said that the use of this tube not only saved the company great expense in tube replacements, but reduced materially the cost of servicing relay stations and added enormously to the dependability of the service.

One member of the committee, intent on ferreting out evidence of monopolistic practices, asked Dr. Jewett whether this tube was available for general use; in other words, whether he could have that tube in his own radio set. Dr. Jewett instantly replied that he thought the tube was made exclusively by Western Electric for use of the American Telephone and Telegraph Company.

The committeeman then inquired, "Has any other company a license to manufacture that tube?" Dr. Jewett replied that he did not know, but thought not.

Other members of the committee, hot in pursuit of monopolistic practices, became interested and asked further questions about the availability of the tube for general use. One asked Dr. Jewett whether he thought the general public should be permitted to buy the tube. This somewhat confused Dr. Jewett, and the episode ended with the clear implication that this valuable development was being withheld from the public use by a great telephone and radio monopoly. This in spite of the courteous efforts of Chairman O'Mahoney to allow Dr. Jewett to refute it.

Because R.C.A. had licensing rights to all of the American Telephone and Telegraph radio patents, they were implicated as a party to this possible monopolistic plot.

Following this episode, Dr. Jewett's testimony proceeded on through the rest of the afternoon. Phil was scheduled to appear at ten o'clock the next morning.

In order to prepare for his testimony the chief of the Patent Department and their attorney asked us to accompany them to the Patent Office. On the way back Dr. Jewett's testimony on the long-life tube was discussed. All felt that his straightforward answers had been greatly distorted and misinterpreted and that it might lead to some unfair implication if the erroneous impression was not corrected. Phil said that he thought he could make the correction quite easily and it probably would come more gracefully from him than from someone connected with R.C.A. or the Bell Laboratories.

We had no sooner arrived at the Patent Office than a phone message came through from R.C.A. in New York asking that they be given an opportunity to be heard on the subject of the long-life tube. They seemed very much disturbed at the possible construction that might be placed on Dr. Jewett's testimony. They were assured that Mr. Farnsworth would straighten the matter out by stating the facts of the case the following day.

Before the conversation ended, however, the R.C.A. representative asked for an opportunity to meet Mr. Farnsworth the next morning.

The anxiety of R.C.A. was not without foundation, because New York papers the following morning carried prominent news articles claiming that a long-life tube was being withheld from the public by a radio monopoly and embroidered the idea quite extensively. The publicity was unjust and unfair, and hardly reflected the temperate inquiry of Senator O'Mahoney and the majority of his committee.

In the morning we met at the Patent Office at nine o'clock to complete preparations for the morning's testimony. Shortly, three R.C.A. representatives appeared with samples of tubes and the R.C.A. Tube Manual showing that some seventy-five types of tubes were available if anyone wished to buy them. There was ample evidence in the tubes, and in the manual, to disprove the implications of monopolistic practice in withholding a valuable tube from public use. It only confirmed what Phil had told us the night before. The R.C.A. men asked Farnsworth to put this material in evidence. I persuaded them that this would not be necessary, that their case would be much stronger if Mr. Farnsworth cleared up the situation in his own way.

As the hearing opened Phil was placed on the witness stand with a battery of three microphones in front of him. I believe this witness chair—"the sounding board of the nation"—with all its impressive physical paraphernalia, gave Phil a bit of stage fright. In all his public appearances he gives the impression of great diffidence. In his own mind he is quite sure of himself, his only anxiety being to state things accurately and correctly so that there may be no possibility of misunderstanding. This gives him the appearance of a halting and hesitant witness until he gets warmed up to his subject.

After the formalities of swearing him in, the Patent Department's examining counsel proceeded to ask Phil questions to clear up the misconstruction placed on Dr. Jewett's testimony. He first asked Phil if he knew of the tube. Phil replied that he was quite familiar with it. The examiner then asked if anyone else, other than Western Electric, could make the tube. Phil said, "We can make this tube if we wish, and I believe that any company licensed under R.C.A. patents is privileged to manufacture it."

Phil then went on to say that we would not manufacture the tube because it was not practical for use in a home radio set and that there were much better tubes available for radio sets that had long enough life for all practical needs. He said that he thought most tube companies had a license to manufacture the tube but, as a matter of fact, he didn't think a license was necessary because he was quite sure that the patent on the tube had expired.

Phil's answers to the questions put to him regarding Dr. Jewett's testimony were so matter-of-fact and simple that they immediately cleared both the American Telephone and Telegraph Company and R.C.A. from the implications of monopolistic practice so far as this tube was concerned. At least the records of the committee were straight, although there was no notice of the correction on the part of the newspapers. It is this sort of injustice that makes large corporations jittery regarding investigations. Scare headlines of monopoly are produced by such misconstructions as were placed on a scientist's answers to questions involving public policy. The records are later set right, but it is difficult to correct the damaging and unfair impression in the public mind.

Following this episode the counsel for the Patent Office proceeded to question Phil regarding his own experience in building up the Farnsworth patent structure and how he had fared in the Patent Office. This was a situation that Phil welcomed. As noted before, serious consideration was being given to a possible revision of the patent procedure and the general operations of the department. It gave Phil an opportunity to defend wholeheartedly the department and its activities. Phil and all of us always had a very high regard for the scrupulous honesty and fairness of the Patent Office. We always have felt that we were given every consideration and had received just decisions in all interferences that developed with our patent applications. Phil's testimony reflected our feelings with sincerity and conviction.

Whenever Phil makes a public appearance he excites great interest. His easy familiarity with scientific terms and phraseologies dazzles and intrigues the listener. This appearance was no exception. He was asked regarding some of his more recent inventions and pulled from a capacious brief case his latest model of an image-amplifier tube, and some of the latest and most interesting of the multipactor tubes. They looked scientifically mystifying in the extreme. He spoke of one of the multipactors as being able to register a candle power of light at a mile's distance. Chairman O'Mahoney and his committee members showed great interest as they gingerly handled the strange devices.

When asked for suggestions regarding possible changes in the patent procedure, Phil reached back into his memory and brought out some painful and disappointing experiences that he had had in his youth by trustingly placing some of his early inventions in the hands of patent attorneys who made a specialty of advertising in amateur publications. He feelingly pointed out that here was an abuse that should be corrected.

Phil had no grievances to register against the department, nor any suggestions of fundamental changes to offer. He seconded some suggestions that had been made by the Patent Department itself in the simplification of the procedure to be followed in patent interferences.

The Commissioner of Patents and his associates were grateful to Farnsworth for his testimony, which reflected so favorably upon the operations of the department. The chief counsel was kind enough to tell us that Farnsworth had covered the situation so well that calling further witnesses in the department's defense seemed unnecessary.

This episode of the testimony before the Monopoly Investigating Committee was typical of occasional public appearances at hearings before government bureaus and commissions. Though Phil enjoyed the resultant publicity, such appearances were distasteful to him and in most cases were not particularly helpful; consequently we tried to avoid his having to spend much time and effort in that direction.

(((25)))

Public Demonstrations in Los Angeles and San Francisco

WHEN THE PLANS for the New York and San Francisco World's Fairs were first announced the expositions seemed to be "naturals" as a launching place for television. There was some discussion of the Farnsworth company having a demonstration at either or both. The Philadelphia plant was so engrossed in advance research work that there was an inclination to put little emphasis upon the demonstration there. However, Bart Molinari, who was carrying on as a lone wolf in our San Francisco laboratories (George Sleeper had joined Dr. Goldmark's staff at the Columbia Broadcasting laboratories), had a determination to set up a practical television unit to show what could be done with the Farnsworth equipment.

As has been indicated previously, Molinari, whom we nicknamed Moli, is a very colorful figure, the son of a wholesale baker in San Francisco. He was in high school during the early days of radio and became one of the very first "hams" to set up a short-wave radio transmission and receiving unit in his home. It was the cherished desire of his father that he finish high school and then go on to Stanford University to graduate in engineering.

Young Molinari had other ideas. During his last year in high school he so frequently played hookey to work on his amateur radio equipment that he lost out on making his entrance requirements to Stanford. While he disappointed his father in this respect, he went on to win distinction and international recognition by capturing the Hoover Cup offered by the then Secretary of Commerce for the best amateur transmitter for short-wave broadcasting.

Moli came to us as a glass blower, but proved to be such a skilled and apt radio technician that we gave him almost as free a hand in development work as was accorded Farnsworth. The back of any radio chassis that Moli built was as precisely constructed as a piece of jewelry. As time went on, almost single-handedly he designed and built up a very practical television camera control and monitor panel. It was beautiful in its simplicity and dependability.

At about the time that this was finished, matters had progressed to a point where it seemed to Mr. McCargar and me that with the help of Kuhn, Loeb & Company we might be able to work out a plan for the expansion of our activities into the manufacturing of radio and television equipment. We recognized that the consummation of any such plan might be a long and expensive undertaking. Since the financing of the company was our responsibility, Mr. McCargar and I felt that during the summer of 1938 we should raise a substantial sum of money to carry us through. As a result I made several trips to Los Angeles. Through these visits I succeeded in providing us with something in excess of $150,000. This, we felt, would carry us through any major underwriting operation necessary to setting up a manufacturing company.

While in southern California, I sensed a possibility of creating some interest in an experimental television broadcasting station using Farnsworth equipment. The heart of the movie center seemed a highly advantageous place for such a station. I therefore made arrangements for an extended showing of Molinari's demonstration unit at a studio adjacent to the old Palomar Dancing Academy on Vermont Avenue. The quarters had large windows overlooking the patio gardens of the Palomar. The whole arrangement was well suited to our purposes.

Our first showing was given to representatives of the Los Angeles press. There was wide interest in television in the film capital; consequently the editors and publishers of the principal papers were present. Tap dancers, a little girl singer, and one or two vaudeville acts constituted the program, which was arranged through the kindness of Carter Wright, a producer associated with Fanchon and Marco.

Among the interested observers were Sid Grauman, Will Rogers, Jr., and others from the motion-picture colony. Moli had all of the equipment tuned to perfection. The weather was unusually hot, and the heat of the lights added to the oppressive temperature in the transmitter room. The program started with a beautiful presentation of the first two acts, then suddenly something went wrong and no picture came through. Moli tried feverishly to find the trouble and get the picture going again, but without success. Evidently something serious had happened. There was nothing to do but tell the audience that a tube had failed and ask their indulgence.

At the time this announcement was made I didn't realize that the heat had cracked the dissector tube. Molinari, with an unerring sense for trouble, found this out very promptly, but it required nearly half an hour to make the change to another tube.

I shall always mark it up to the everlasting credit of the publishers, editors, and feature writers of the Los Angeles papers that in the news story that appeared the next day we had headlines announcing the first public showing of Farnsworth television in Los Angeles and saying it was "just like the movies" —the highest praise that could be given in the movie capital. Not one word was said regarding the embarrassing broken tube and delay in the program!

At the Palomar Studios next door one of the nationally famous orchestras played every evening and a crowd of young people came to enjoy the dancing. During intermission they would gather out on the patio. Looking through the window, Moli and I would often pick out what we considered good television subjects and then go out and ask them to come in and be televised. We got to know what skin textures seemed to televise best and what type of hair gave the most pleasing results. In fact, we used to place bets with others around the studio as to whether this one or that one would really show up well on the television screen.

One afternoon I was invited to a social gathering. Later the guests were to go to the studio to see a television showing. As we left for the studio I asked one of the guests if she would be kind enough to be televised. She graciously consented. Later I learned that she was Billie Dove of the movies. Of all the people that I have seen televised I think that she was the most perfect subject. Among the men, John Boles was the best. Both of them not only had the beautiful regularity of features that shows up well on a television screen, but they also had the proper texture of skin to give best results.

Young Will Rogers, Jr., was fascinated with our developments, and Molinari arranged a special showing for him and his mother and family.

It was in Los Angeles that we first televised a hula dancer, Princess Louana, a native Hawaiian who performed at many of our showings. To me it was the most successful of all our acts.

We continued the television demonstrations for a period of five weeks. Many of the famous figures in motion pictures took advantage of the showing. One evening was given over to a demonstration for Harold Lloyd and his family. When he was televised one of his children said he looked pretty tough. He had appeared without make-up and his beard showed through very badly, as is often the case with men unless the beard is covered with make-up.

It was here that we had our first real test regarding television make-up. The Max Factor Company of Hollywood was anxious to learn something about the requirements of television to prepare themselves to do the same splendid job in television that had characterized their work in motion pictures. Max Firestein, head of the Factor Company, spent a great deal of time at the studio. In addition, his nephew, Sidney Cramer, was assigned to make up the television artists as they appeared at all of our showings. We gave two demonstrations a day, one in the afternoon and one in the evening. Cramer usually arrived at the studio each day with some new mixture of coloring to try out. He finally found that the best effect to be had on the lips was not to paint them blue as we did in the first showings, but to mix a little blue pigment with the ordinary type of lipstick. This gave a more natural shading to the lips than the blue make-up.

We could do wonders with most television subjects by painting out lines and wrinkles with liberal use of red make-up. One evening when a leading moving-picture producer and his wife, a former Follies beauty, came to one of our demonstrations, I asked the producer's wife if she would like to be televised. She was well in her forties, and while the lines of her former beauty were still evident, her features were a bit heavy and there were distinct crow's-feet in her face. But with the skillful use of blue make-up under the chin and red lines on her cheeks and around the eyes, we succeeded in transmitting a television picture of her former loveliness.

The producer, who was something of a wag, watching at the receiver, said, "My God, is that my wife?" I assured him that it was. He asked, "How much do these machines cost? I think I'll buy one if it can do that much for my wife's looks."

At this Los Angeles showing a representative of Scientific Films, Inc., which did short news subjects of a scientific nature for release through Paramount, became interested in producing a film of the Farnsworth television development. Arrangements were made for Phil to be photographed at our laboratories in the east, and later the film was completed in San Francisco. This was the type of public recognition that Phil enjoyed most, and he took great pains to co-operate fully in order that the film should give a true representation of our development. The picture was taken in Technicolor and given worldwide release.

On the whole our showing in the motion-picture center was kindly received, although in some quarters there was still a lurking fear that television would be a serious rival to motion pictures.

Following the Los Angeles show Molinari and the equipment returned to San Francisco. Later an extended showing was made under the auspices of the San Francisco Merchants Exchange. This was a more finished effort than the Los Angeles demonstration. Molinari had learned better lighting technique and handled the television camera to better advantage. Here again the Max Factor Company assigned Mr. Cramer to care for our make-up problem, and Carter Wright again provided the program. One of the most successful acts was a classic dancer.

These two showings by Molinari were the first real attempts at such extended performances of television for the public. R.C.A. was broadcasting programs over the experimental station in New York, but there was no scheduled program where the general public could see both transmission and reception. Don Lee, in Los Angeles, under the direction of Harry Lubcke, was also doing considerable television broadcasting; however, they would not permit anyone to see their transmitting equipment. It was not until after we had completed our public showing that Don Lee adopted the wholly electronic method of transmission.

The following season the San Francisco and New York World's Fairs opened, but we did not feel that we could afford to put in a television demonstration at either fair.

(((26)))

Manufacturing Plans Completed

In surveying the field to get the best-qualified man to head the proposed manufacturing organization of the Farnsworth company we centered finally on E. A. Nicholas, head of the Licensing Division of R.C.A. A tentative commitment was secured from him to accept the presidency of the company in case the plans under consideration materialized.

Here at last Phil's dream of a large, adequately financed television company was taking tangible form. Mr. Nicholas was regarded as one of the most able and well-liked men in the entire radio industry. He was young and aggressive and had courage and imagination. That he was willing to leave his splendid association at R.C.A. and take his chances with us was to us added proof of the soundness and value of the Farnsworth developments.

As soon as Mr. Nicholas entered the picture he set about energetically to shape up a commercial organization with the soundness and potentialities that would attract adequate capital to launch the new manufacturing company.

It was learned that a radio manufacturing company, the Capehart Company, in Fort Wayne, Indiana, could be purchased and that another near-by plant at Marion, fifty miles south of Fort Wayne, was also available. The upshot of it all was that a plan was finally worked out for the acquisition of these manufacturing facilities and for the floating of $3,000,000 worth of Farnsworth stock to provide for the purchase price of the two plants and to insure adequate operating capital.

Our goal was to have the arrangements completed and the new company in operation some time in November 1938, if we could get the matter cleared through the Securities and Exchange Commission by that time. Since it was a perfectly legitimate speculative undertaking it seemed to us that there should not be a great deal of red tape involved. However, in this we were mistaken.

Naturally we engaged competent attorneys and accountants. As we got into the necessary work of filing the registration statement we recognized that a real task lay ahead. While our bankers expected that there would be some routine effort to get the matter through the newly set-up regulations of the S.E.C., we were wholly unprepared for the difficulties we met at the hands of this government bureau. We were undoubtedly the victims of a situation that had been created by flagrant abuses of buyers' confidence in public financing. The position taken by the Commission seemed to be that you were guilty until you proved your innocence. In our case the attitude seemed to be one of daring us to prove that we were not all a pack of dishonest rogues. The expense that we were forced to incur mounted to a sum several times our most liberal estimates, and the demands of the Commission required seemingly endless trips of accountants, lawyers, and executives to Washington.

The registration statement and its supplement, in final printed form, represented a weight in printed matter greater than that of the New York telephone directory. Upon any point under discussion expert after expert would be brought in by the Commission to question our attorneys and accountants.

Jesse McCargar and I were held in New York for months doing what we could to help things along. Since we were on one side and Kuhn, Loeb and the underwriters were on the other, we had to bear the expense without any help from the financial interests until the underwriting was approved. The two of us and Phil were worried that the delays would make the underwriting impossible. This was during the ill-fated period of appeasement of the Germans by Prime Minister Chamberlain. The success of any underwriting depended on the action of the stock market. The market in turn was acting with a disconcerting uncertainty that grew worse as the precious days went by.

Our experiences were not only disconcerting but sometimes amusing. One instance was typical. Our chief attorney, John F. Wharton, and some other of our attorneys and accountants, were discussing some item in our financial statement with the commissioner's staff. An impasse arose on the particular item and a Commission accountant said, "We will call in our expert chief accountant on this matter."

With that, a youngster, not long out of college, made an impressive entrance into the room, sat down at the table, pulled a comb out of his pocket and smoothed back his luminously dressed hair. Mr. Wharton modestly said that he felt at a loss in discussing the item in question with a qualified expert, and then went on to prove that our contention was right. After Mr. Wharton had established his point he didn't take the trouble to inform the youthful authority that he was the author of a textbook on legal accounting. The young expert retired in good order, with his hair unruffled.

Time and again our attorneys returned from Washington feeling sure that when certain demands and recommendations were complied with our clearance would be granted, only to be confronted with further delays and stipulations that had to be met.

As it was, the issue was put on the market on the very last day on which it would have been possible to consummate the underwriting. We finally got our release from the Commission and there were several days of frightening uncertainty in the market during which Mr. McCargar and I could readily see the whole structure crashing down around our heads.

Fortunately the market stiffened, and March 31 was set as the closing date. Representatives of all the parties concerned, with their attorneys and accountants, assembled at the appointed hour in the directors' room of the offices of the chief underwriter. Everything was prepared; there were 113 legal documents to be finally ratified. The attorneys were physically and nervously exhausted as a result of the long siege. We were tense with the fear that some last-minute failure on someone's part might upset the whole thing. The bankers were anxious because of the very uncertain market conditions. Mr. McCargar and I therefore breathed a great sigh of relief when the check for $3,000,000 was formally handed to me as treasurer of the company and I was able to dispatch it to our bank for deposit.

On the following day Hitler invaded Czechoslovakia. Had we been delayed one more day the underwriters would of necessity have had to withdraw.

Thus the dream of the launching of the Farnsworth Television and Radio Corporation as a commercial organization finally came into reality.

Mr. Nicholas and I proceeded at once to Fort Wayne. The plan of action provided that the Farnsworth company should engage in the manufacture of radio receiving sets to carry the company along until such time as television was launched commercially.

On the first of April 1939, Mr. Nicholas, as president, had two well-equipped plants and operating capital as a nucleus with which to get the business under way. With amazing speed a manufacturing and distributing organization was built up. By the middle of August of that year the company had a complete line of radios ready for the market and 75 per cent of the sales territory of the United States covered with excellent distribution. At our first distributors' conference approximately $1,000,000 worth of radios were sold. Thus the manufacturing business was launched.

In the meantime plans were under way to transfer our Philadelphia research laboratories to the Fort Wayne plant. This was effected later in the fall.

Thus Phil Farnsworth's dream of facilities for the commercialization of his television plan became a reality. Ill health and a strange hesitancy prevented him from stepping immediately into the new setup. The manufacturing plants had been in operation for some months before he finally moved from Philadelphia to Fort Wayne. Everything was done to provide him and the company with the laboratory facilities that he had always dreamed of but had not been able to get in such completeness.

He was made a director of the new company and had a place in the management group as vice-president in charge of research. An appropriate office was provided, and his salary was fixed on a par with the others of the management group.

In setting up the company, Mr. McCargar and I had protected his interests in stock holdings to the extent that when the new company came into being Phil was the largest individual stockholder. In working things out this way we took great satisfaction in the fact that here for once an inventor had come out of a long-drawn-out promotional struggle with more stock holdings than any other individual in the company.

Upon arrival at Fort Wayne, Phil pitched into the development work with his old zest and fervor. He was not content to confine himself to research alone, but became interested in every phase of the operations of the new venture. Then one day he walked into the office of President Nicholas and announced quietly, "I'm getting out."

A slow smile spread over his face when he saw the look of alarm from Mr. Nicholas. "I'm just going fishing," Phil added.

The truth was that Farnsworth had been driving himself so hard he was endangering his health. He was beginning to show the strain of his many hours of daily work, and doctors had advised him that he would have to slow down. In addition, Phil was not fully satisfied with his new job because it required him to do so much planning and organizational work, necessitating conferences and routine checking. He longed to spend all of his available time in the research laboratory, where he could do the inventive work himself rather than merely oversee it.

Officials of the company agreed with Phil that he could continue to do the creative work he loved so much, and at the same time safeguard his health, by moving to his Maine farm. There his own laboratory was installed—complete with the tubes, coils, and other weird-looking electrical devices necessary to enable him to conduct all types of development work. He continued to spend considerable time in this private laboratory, but he also had within walking distance his own private lake where he could fish to his heart's content, and wooded hills where he could walk by himself and muse over the experiments he was conducting.

Only occasionally did Farnsworth return to the company's headquarters plant in Fort Wayne, but he kept the research engineers there well supplied with ideas for new inventions and ways of perfecting earlier discoveries. His wife still worked with him at times in his laboratory and discussed with him some of the perplexing problems of his work. And at times he got advice —not always solicited—from his two teen-aged sons.

((27))

R.C.A. Licenses

In his dream of becoming a great inventor, Farnsworth was continually aware that patent protection and patent recognition were the keys to his scientific and financial success. Royalties from manufacturers licensed under his inventions were looked to as a material incentive for his work and a means of financing further research. His patent structure was designed to cover his inventions in the whole field of television. At the same time, he was realistic enough to know that others would be working to establish strong patent positions, and he anticipated that Radio Corporation, with its vast resources, would be the chief contender.

Farnsworth had begun to apply for television patents even before he was old enough to vote. By the time he reached the age at which most young men are finishing college and preparing to start careers in the business and scientific worlds, Farnsworth had been granted patents covering his system of transmitting and receiving television images.

As the youthful inventor continued his developmental and research work during the late 1920's and early 1930's, the patents began to come in steadily increasing volume. A whole nest of patent claims developed around the construction of the dissector tube. Another group of patents resulted from his refinements of the scanning wave. Many patents were secured from work on amplifiers, and his research in the projection type of receiver tubes resulted in the filing of many additional applications. More than a hundred television patents have already been issued to Phil Farnsworth, and scores of others have been granted to his associates who have worked with him through the years. Many more cases are pending in the Patent Office.

The rapidly growing patent portfolio represented completed goods on the shelf to be leased to the best possible advantage. During the long anxious years while this accumulation was taking place, there was occasional uncertainty in Phil's mind as to how firm a place he could secure in the new art to which he had contributed so much. Some of his patents and allowed claims in unissued cases seemed to assure him of the prominence he deserved, but for many years while some of the more important ones were being fought for in interference proceedings, there was always a chance that he might lose some of the claims.

As his position became stronger, we began to consider ways to develop an interest in our patents. As already noted, the first contract was signed with the Philco Company. The European licenses and the cross-licensing arrangement with the American Telephone & Telegraph Company gave added strength and prestige to Farnsworth's ever-increasing patent portfolio.

Because of its strong position in the radio industry, R.C.A. had control of many of the major patents in that art. Manufacturers of radio receivers and transmitters operating under license from R.C.A. were paying royalties of several million dollars annually to that company. R.C.A. had also become our chief competitor in television, so it was a foregone conclusion that if Phil's patent structure were to be placed on a sound commercial footing a licensing arrangement would have to be negotiated with the Radio Corporation.

His backers not only had to establish Phil's claims over R.C.A. in important interferences in the Patent Office, but it was necessary to obtain a licensing arrangement that would result in the payment of substantial royalties to the Farnsworth laboratories. The terms of such an agreement would determine the measure of success of our venture.

R.C.A. first took serious notice of Farnsworth when the Philco contract was being concluded. David Sarnoff, R.C.A.'s president, had visited our San Francisco laboratory and discussed the possibilities of a patent agreement in a very general way, but for some time we heard nothing further from him.

As time went on, sporadic attempts were made to establish a basis of negotiations. They usually resulted in vague suggestions and delay. This delay was annoying, but Mr. McCargar counseled us to have patience. He quoted Mr. Fagan as often saying, "If you lean against anyone long enough, he will move over."

As television came nearer and nearer to a commercial reality and as Farnsworth's patent strength was increased by the winning of several important interferences, R.C.A. recognized that some arrangement had to be made to enable them to use the Farnsworth inventions. After Kuhn, Loeb & Company came into the picture, Hugh Knowlton assisted in putting negotiations on a more realistic and formal basis and stepped up the tempo of the discussions. At first, the two sides seemed so far apart that it looked utterly hopeless. Only the clear underlying fact that neither company could get along without the other kept the discussions alive.

To Phil these were months of uncertainty and nerve-racking

anxiety. After spending thirteen years in building a patent structure, he faced the all-important question of whether the leader among the possible customers, who would set the pattern for all the rest, would agree to pay for it. An equally serious problem was the question of whether the price would be adequate. Phil, Jess McCargar, Don Lippincott, and I often had long discussions regarding the matter. We felt convinced that an arrangement must be made, but we too had many anxious moments and some misgivings as to the type of deal that could be negotiated.

Previous to this time, the Radio Corporation had never paid continuing royalties to anyone. It had succeeded in purchasing any patents that were necessary for its operations, but we were not in a mind to sell. Neither were we experienced in patent negotiating of this sort. Consequently, we drew a sigh of relief when, with the launching of the new company, experienced people entered the picture.

Mr. Nicholas, the new Farnsworth president, was an old hand at negotiating. He had for many years been in charge of the Patent Licensing Division of Radio Corporation. He was ably seconded by Edwin M. Martin, directing head of the patent division of the new company. Mr. Martin had previously been with the Patent Department of the Hazeltine Corporation. Both these men recognized the importance of renewing licensing negotiations with Radio Corporation, and long experience had taught them how to go about it. They knew the fundamental strength not only of Farnsworth but of Radio Corporation as well. Therefore they could talk on equal terms with R.C.A.'s Mr. Schairer, vice-president in charge of patents.

The negotiations were not easy, but the two Farnsworth men were determined to make a fair and equitable arrangement on the part of both companies. Of course, Mr. Schairer preferred to purchase patents outright rather than enter into an arrangement where continuing royalties would be paid. Mr. Nicholas and Mr. Martin were equally certain the patents were not to be sold.

Conversations and negotiations continued from May to September of 1939. Finally an agreement was reached and a last dramatic session was arranged. At five o'clock in the afternoon, after an exhausting three days of whipping the agreement into final shape, Mr. Schairer led a group of top Radio Corporation officials into the conference room in Radio City. The contract in its final form was brought in. All of the men were tired, but all were pleased that a satisfactory arrangement for both companies had been reached. When Mr. Schairer finally signed the agreement, there were tears in his eyes. It was the first time that his signature had been placed on a contract whereby the Radio Corporation had to pay continuing royalties for the use of patents.

None of the three persons who had the most vital interest in the completion of this arrangement—namely, Phil, Mr. McCargar, and I—was present when the agreement was concluded. Mr. McCargar was in New York, Phil was in Maine, and I was in San Francisco.

Immediately upon signing the contract Nicholas and Martin called Mr. McCargar at the Plaza Hotel. He told them that the event called for a celebration and asked them to join him and Mrs. McCargar for dinner in the Persian Room. It was a gala evening. Toasts were drunk and wires of congratulation were sent to Phil and me. Though we were not all together to celebrate the success, each knew that the others rejoiced in the victory that had firmly established the Farnsworth inventions as vital patents in the field of television. It was a completion of thirteen years of struggle and effort in which Phil, Mr. McCargar, and I had worked together for a common purpose. Mr. Nicholas and Mr. Martin had succeeded in putting an important seal of recognition on Farnsworth's inventive genius and the patience and persistence of his backers.

(((28)))

Television Standards Agreed Upon

THE FARNSWORTH Television and Radio Corporation was formed around the television inventions of Farnsworth. To make money and keep the company going it was recognized that until television came into full commercial use the company bread and butter must be made from the manufacture of radio sets. This, however, did not divert them from the main purpose of getting television launched commercially as soon as possible. In this purpose Mr. Nicholas and the management group were beset with many difficulties and discouragements.

The attitude of the Federal Communications Commission and the general feeling throughout the radio industry was that with the opening of the year 1940 television would be launched on a commercial basis. From the point of view of timing, therefore, we were in good position to take advantage of the market that we were sure would develop during 1940 for television receiving sets.

Television for the public seemingly was to be born, like Venus, full-grown—or, as Phil once put it, "The baby is being born with a full beard." That has been the pattern which television has followed. Several factors contributed to this. First among them, we were building a new industry that in the minds of some would be in conflict with two existing industries, motion pictures and radio. In fact, the motion-picture industry at one time seriously considered taking effective steps to boycott television. A meeting was called by the motion-picture moguls in New York to devise ways and means to accomplish it. Legend has it that Walt Disney's company refused to go along with any such strangling tactics, and the plan fell through.

For several years after television had arrived at a stage where it would have been possible to launch it commercially, the radio industry was enjoying prosperous business. Some companies, therefore, did not wish to disturb a good thing for some untried project that would require huge capital expenditures and the development of a new market. At that time there were honest and serious misgivings on the part of some of the radio leaders as to the future of television. Many doubted its value as a medium of advertising. Others felt that program costs would be so high that the operation of a television station could not be profitable. Farnsworth and his backers were practically the only ones doing television experimental work who did not have conflicting interests in radio and whose whole heart and interest were in the commercial exploitation of the new art.

Some elements in the radio industry were openly antagonistic. One prominent manufacturer put out a folder attacking the efforts to introduce television as a commercial product. This was widely distributed among radio dealers and technicians. Others emphasized the great expense of transmitters and receiver sets as a deterrent to wide sale and use of the new product. There was little to be done to counteract such propaganda, other than to keep everlastingly at improving the picture and simplifying the apparatus.

It was a discouraging fight for Phil and his backers to meet the efforts of some groups entrenched in profitable radio business to delay and hamper the introduction of commercial television. The struggle was the more difficult because of the intangible, though effective, nature of the propaganda.

During those years we received considerable encouragement from the Radio Commission, later known as the Federal Communications Commission, through the granting of radio channels for the television experimental work. Recognition of the new development took tangible form when the F.C.C. requested the Radio Manufacturers Association to recommend standards for commercial television.

The amateur "ham" operators who had been so helpful to radio in its early years were entirely excluded from any active part in the development of television and from making any real contribution to it. In the first place, equipment was so expensive that it was beyond the reach of most amateurs. Also the technique of building a television camera and transmitter required an order of engineering knowledge beyond the abilities of most of them. An amateur might possibly have made an operable receiving set, but there were only two places, New York and Los Angeles, where any television programs were to be had.

As the Committee on Standards of the Radio Manufacturers Association went on with its work of developing adequate standards for television broadcast to recommend to the F.C.C., there developed within the Commission a more rigid position toward its commercial exploitation. The encouragement we had earlier received appeared to dwindle. With unemployment as a major national problem it seemed to us that through the launching of television, opportunities might be furnished for large numbers of people in the manufacture of a new instrument for home enjoyment. The Commission seemed to direct its efforts toward protecting the public from buying an article that had not been developed to its ultimate perfection.

In the meantime the radio business fell off to some extent and the industry as a whole, with but minor exceptions, was in favor of aggressive action in television.

The report of the Committee on Standards was finally finished and approved and submitted to Washington in the fall of 1939. At this time the National Broadcasting Company was doing some good television broadcasting of plays, news events, and studio programs, and a limited number of television receivers had been sold to the public, principally in the New York area.

The Columbia Broadcasting System had acquired the Chrysler Tower as an outlet for their programs and was setting up equipment and studios in anticipation of the launching of commercial television broadcasting.

Months went by without any action from Washington on the Radio Manufacturers Association's standards. Finally David Sarnoff, taking the leadership for the industry, reported to the F.C.C. that his company had spent $10,000,000 on television development and others had also spent large sums for the same purpose, and he urged the Commission to take some action. Sarnoff felt that he could not justify such vast expenditures with his stockholders unless something concrete in the way of commercial returns were forthcoming in the near future. The Farnsworth company took the same position.

After this it seemed the way was cleared for reasonably prompt action on the part of the Commission, since its Committee on Television had made a favorable report on the standards submitted by the Radio Manufacturers Committee and had recommended that limited commercialization be permitted. Then, during what was expected to be a routine hearing before the Commission, there was injected a hypothetical possibility of standards being changed within a short time, thereby causing any television sets sold to become obsolete. Although there was no likelihood of anything of this nature on the horizon, this led to other disagreements within the membership of the R.M.A. Finally, after extended public hearings, the Commission smoothed out the differences and ruled that the adoption of standards would be deferred, but that limited commercial television broadcasting would be permitted commencing September 1, 1940.

Eager to forge ahead in its development program, one company immediately placed an advertisement in the New York newspapers announcing the doubling of their experimental broadcast periods and a reduction of the sales price of television receiving sets by one-third.

The F.C.C., with the exception of one official, took offense at what seemed a reasonable and progressive action on the part of this particular company, promptly delivered a stinging rebuke to the company for its activity and accompanied it with an order rescinding the ruling permitting commercial television after September 1, 1940.

After some time the Commission asked the R.M.A. to appoint a new Committee on Standards, known as the National Television System Committee, to re-evaluate the standards for television broadcasting. The R.M.A. appointed a large and representative committee of engineers with Dr. W. R. G. Baker, of the General Electric Company, as its head. Upon organization it was divided into nine panels, or subcommittees. Farnsworth had a representative as chairman of one panel, and representation on seven others.

After a year's careful study the report was finished and filed with the Commission. It consisted of several volumes, one for the findings of each panel. The work of the original Standards Committee was so well done that Dr. Baker's committee found little to alter. The major recommendation changed the number of lines in the television image from 441 to 525.

The Commission arranged a hearing on Thursday and Friday, March 20 and 21, 1941, to consider the report of the National Television System Committee. The sessions continued on through the following Monday. All who testified before the Commission were in agreement that the standards recommended by Dr. Baker's committee should be adopted. The majority favored immediate commercialization.

On May 2, 1941, Commission Chairman Fly announced the adoption of the standards recommended by the R.M.A. Committee and announced that full commercial licenses would be granted qualified applicants beginning July 1.

(((29)))

At Long Last—Commercial Television

EVERYTHING seemed ready for the launching of television in the summer of 1941. At long last, commercial television had become a reality, fifteen years after Phil began his first experimental work in his little apartment in Los Angeles.

Mr. Nicholas and his staff were gearing the Farnsworth company to launch into a manufacturing program. There were, however, some misgivings because of the international situation. These soon became realities. The need for production of war materials became so imperative that on April 1, 1941, a general order went out to the industry that manufacturing of commercial radio sets was to be discontinued. This meant that television was indefinitely delayed. War seemed imminent. If it came, commercial television would have to wait until peace was here again. As a consequence, all of the Farnsworth plans had to be revised to meet the new order.

The company's management met the challenge by going aggressively into the development and manufacturing of vitally needed electronic equipment. Because the company had no history of contracts with the War or Navy Departments, direct contracts were not immediately available. However, its management had the confidence of such organizations as General Electric, Radio Corporation and Western Electric. These companies were overloaded with war work. As a consequence, the production lines of Farnsworth were soon filled with subcontracts from the older organizations which were working on direct government contracts. The work was executed so economically that it was not long before all the facilities of the Fort Wayne and Marion plants were running at full capacity on government work, most of which was from direct contracts.

Because the original Farnsworth company's activities were entirely in the field of research, the new company had established a large and well-manned research department. The laboratory itself was completely air-conditioned and excellently equipped for electronic research. All of the facilities and skilled personnel were immediately put to work on research problems having to do with the war effort.

In this field the Farnsworth dissector tube was singularly useful. Its principle was used most successfully in the sniperscope, an infrared device which enabled our soldiers to see in the dark without being seen by the enemy, and in the development of other useful military devices. A special wing was added to the Fort Wayne plant to provide facilities for the manufacture of specialized tubes. A large section of this was given over to the manufacture of a small dissector-type tube used in the sniperscope. Here a corps of women was trained to do specialized work on the tubes. The room in which they worked was as immaculate as a hospital ward. Another room was utilized for production of dissector tubes for other war uses which even now are not widely publicized for security reasons.

The extensive production lines at the Marion plant were converted to the manufacture of a wide variety of electronic equipment for Army and Navy use. At one time 85 per cent of all the transceivers for tanks were produced here. Other lines were given over to radar and radio equipment for planes, and still others were set up for production of transmitters and receivers for different types of naval craft, particularly PT boats. Night and day, month in and month out, through the war years there constantly flowed beautifully engineered radio and radar sets from these well-organized mass-production lines.

There was such a demand for the products Farnsworth was putting out that in order to find additional personnel and facilities for expanding production, a former furniture plant in Bluffton, Indiana, was leased and lines were set up there to build the smaller electronic units. Before long, this thriving community added five hundred employees for work on the Farnsworth production lines. Along with this the research laboratory at Fort Wayne was working on new developments needed by the War and Navy Departments. These activities were carried on under strict wartime secrecy. Even today some of them are still under wraps. In this work there was the closest co-operation between the Farnsworth company, Radio Corporation, Massachusetts Institute of Technology, and the laboratories at Wright Field and other important development centers. In this effort, the results of Phil Farnsworth's genius were being utilized fully in the prosecution of the war.

Early in the war, I visited Phil in Fryeburg and found him in a bad state of mind and health. He was fretting under the doctor's ban on concentrated work at a time when he felt the country needed all that he could give in the way of new inventions and developments. His doctor was adamant in ordering him not to work too much. In spite of repeated warnings, Phil, in his laboratory in Maine, did overtax his strength by working long, irregular hours. Although he was losing weight and growing weaker, he persisted in spending much of his time in his laboratory. Finally he yielded to the doctor's urging and he went to a hospital in Portland, Maine. His condition became so bad that for many weeks while in the hospital there was little hope for his recovery. However, the will to live and the inheritance from his tough Mormon ancestry gained the upper hand.

Finally he returned home in a very weakened condition. For months it was necessary for him to be in a wheelchair when he was out of bed.

During this period I visited him and admired his stubborn spirit for survival. It was then he had his first meal at the dining table with his family. It was a gala affair. Phil and his boys, in co-operation with the state game commission, had made an effort to raise specimens of wild life that might be released and eventually become indigenous to the countryside. For this purpose they had a pen of wild pheasants breeding on the place in a separate enclosure. For this dinner the boys had taken some of the pheasants and Pem had cooked them to perfection. Phil, for him, ate quite heartily, and it was an occasion of great happiness for all concerned. He was again in the family circle.

While I was there he talked of the things he would like to be doing for the war effort, but his doctor had told him to take it easy. He was able to devote only about an hour a day to work on the things he had on his mind.

To divert his attention from the more taxing aspects of the electronics field, he helped his brothers organize a company to cut some hardwood lumber that was needed for the boxing of war instruments. This turned out quite successfully and was a diversion rather than a burden. Carl and Lincoln, the brothers, took over the responsibility for the development of the plant and production of the much-needed lumber.

During the war, from time to time, Phil would get too hasty

in his eagerness to return to his laboratory work and would become so ill that he had to return to bed. The general report on his health, however, was a steady but very, very slow improvement. It would have been a great help to the laboratory in Fort Wayne to have his personal advice and help during these war years, but it was not possible to ask it.

Late in 1942, in order to be most useful in the war effort, I joined the staff of the Radiation Laboratory under the direction of Dr. Ernest O. Lawrence at the University of California in Berkeley. This was a part of the atomic bomb program. The need for haste in this project was such that I had little time to devote to Farnsworth matters beyond attending occasional directors' meetings. Therefore, it was not possible for me to take time out to visit Phil again in Maine. I did occasionally hear through Mr. Martin, who was handling patents for the Farnsworth company, and others of Phil's progress. I was bothered not to be able to go to see him and get a better understanding of his physical condition.

In the meantime, the work of the Farnsworth company went on at a pace that was building a splendid reputation for the production of war materials. Under the inspiring and capable leadership of Mr. Nicholas and Mr. Martin, the morale of the company was tops. Production per individual was high. Everyone was jubilant over the first E awarded for excellency in war production by the Navy Department. This was followed by similar awards again and again until the company had accumulated seven for the two plants at Marion and Fort Wayne. The former came off with the larger number of them, due largely to the fact that the production lines were permitted to run a longer time on one type of electronic equipment. From all areas of the war fronts came word that the equipment built by the Farnsworth company was doing its part in the far-flung efforts to win the war.

It was a source of great disappointment to the directors not to have Phil at our meetings. The laboratory staff also missed his leadership and inspiration. However, all understood his inability to be with us and looked forward to the day when he would return.

When the war ended there was a considerable period of readjustment to civilian production. Shortages of all kinds hampered activities, but soon the company got back into its stride of producing radios and phonograph-radios, for which it has become nationally famous.

Word came from Maine that Phil's health was very much improved and that he had bought an airfield. He was taking an interest in aviation for a diversion. He and his brothers had formed a little company and had purchased planes for use at their field. Carl and Lincoln were on a trip to pick up a plane when they were overtaken by a storm and were forced to make an emergency landing. In this crash landing, Carl, who had been the mainstay and had helped Phil in all of his enterprises, was killed. This was a severe blow, but fortunately it did not retard Phil's general slow return to normal health.

Some time later, after a long period of absence, it was with pleasure that the Board of Directors welcomed Phil to one of its meetings in New York. He was still thin and his face bore the marks of physical suffering, but the spark of genius seemed to be burning with the old glow, though tempered with a more mature judgment.

At the first two meetings Phil attended, I was not present. Therefore, when I went to the monthly meeting in October 1947, I was disappointed that Phil did not appear for it. He had been expected; Mr. Nicholas told me Phil was planning to be there. In one morning's deliberation it seemed imperative that we get Phil's approval for something the directors wanted to undertake, so a telephone call was made to his Maine home. Word came back that Phil was cruising in a plane over the area that was burning in a disastrous forest fire. It was the area immediately around Phil's home.

A couple of hours later Phil called back and told us that because the whole countryside was in flames, it had not been possible for him to come down to the meeting. He was calm and factual in his discussion of our business matters and in his report of the fire. He said he thought his property was safe. It was not until two weeks later that I learned in a letter from Pem of the disaster that overtook them. She said that while Phil was talking to me the flames were coming toward them and within fifteen minutes struck the home. Within a half hour all of their buildings were destroyed. Later, when I saw them, Pem and Phil told me of the incredible swiftness with which the fire had swept in upon them. Fortunately, they had planned to spend the winter in Newton Center, Massachusetts, and had taken a house in that community. They had moved some of their household belongings out of the Maine house for shipment to Newton Center. Most of Phil's prized books and mementos of his long years of research were a complete loss, but Phil and Pem took the loss in their stride.

(((30)))

Research Goes On

MY FIRST MEETING with Farnsworth after the war was at a directors' session in Fort Wayne in February of 1948. On this visit to Fort Wayne Phil renewed his work on a projection-type tube for television receiver sets. The engineers set to work in earnest in a final effort to lick this baffling problem. I spent a couple of hours discussing all aspects of the research with Phil and the engineers. It was a continuation of the age-old problem that seemed insoluble—that of getting sufficient intensity of light on each scanning point on the small receiver screen to permit its projection on a large screen with sufficient light intensity to give a good image. It was like old times in the Philadelphia laboratory to hear the discussions of what might be accomplished.

Throughout our research development Farnsworth has always had some work under way in an effort to get what he called a projection-type receiver tube. Light intensity on the small screen was the problem. There seemed to be no way to add to the light emanating from the fluorescent screen and still maintain the varying intensities making up the image. As far back as 1931 Phil had operating in our laboratory a projection-type cathode-ray tube that would throw an image three feet square on a screen. It had a willemite fluorescent surface.

Such a picture was shown to David Sarnoff at the time he visited our laboratories. He was so impressed with it that he asked me why we showed the smaller image at the end of the large receiving tube. I told him that while the projected image was quite satisfactory, we had to use such high voltage in producing the scanning beam that the fluorescent surface did not have sufficient life to make it commercial. There was also the expense of lenses and the problem of focal lengths to give proper results on a screen built into the cabinet.

Phil had done extensive experimentation with different types of screens. One screen he used showed great promise. It was composed of hairlike tungsten wire woven to a fabulously fine mesh. In order to produce the mesh, we had purchased our own weaving machine for the experimental work. Harry Bamford, formerly of our research staff, spent considerably more than a year's time working on this device.

The receiving screen enclosed within a vacuum tube was of about postage-stamp size. It was so constructed that it could be heated to a red glow for operation, then when scanned by the cathode beam, the cathode ray would heat the scanned points to incandescence. The major difficulty in the operation of the device was the rapid extinguishing of the incandescent point immediately after scanning. Bamford finally solved the problem of weaving the fantastically fine mesh and succeeded quite well in overcoming the extinguishing problem, but in spite of the months of painstaking effort it was never brought to a wholly successful conclusion.

While Farnsworth's experimental work on miniature screens for projected images was going forward, the Germans in the Fernseh Company succeeded in perfecting a willemite fluorescent surface that would stand bombardment of high voltage. As a result they were able to project a television image on a screen of the size used in motion-picture theaters. However, the apparatus used in getting this large image was so expensive that it is not usable except in large auditoriums.

We discussed all these matters with the engineers, and this was the first of repeated trips at more frequent intervals that Phil made from his home in Massachusetts to the laboratory. The doctor had given him the go-ahead sign, and his health was improving by rapid strides. He ate with relish and was making slow progress in gaining some weight. It was now possible for him to attend all of the directors' meetings. The doctor's permission for full steam ahead again acted like a tonic to his health. During the years of his desperate illness and slow convalescence, he had learned that rest is essential if one is to do his best work. So in spite of his eagerness he took his work at a pace which he could stand.

A few weeks after this meeting in Fort Wayne, I was delighted to receive a telegram from Phil saying that he and Pem were arriving in San Francisco to visit their families and that he wished to spend two days with me at the Radiation Laboratory at the University of California, where I had charge of scientific personnel. When he arrived, he telephoned and with some diffidence asked if he might bring his sister and Pem to see the large cyclotron on the hill. This was arranged and they were shown the huge installations. Here Phil was in his element. After the sight-seeing trip was over, Pem and Phil's sister left and I introduced Phil to Professor Lawrence and other leaders of research on the staff. Since much of the work here is in the field of electronics, Phil's eager mind was grasping and searching out the essentials of the work. Because of Phil's specialized knowledge in electronics, Professor Lawrence, the director of the laboratory, asked him to become a consultant without fee.

After the visit here, he and Pem went on to see friends and relatives in St. George, Provo, and Salt Lake City in Utah. In all of these places they were given impromptu receptions. This trip did Phil a tremendous amount of good. It was sort of a release from the bondage of his long illness and the frustration of enforced inactivity.

Immediately upon his return to Fort Wayne, he announced to Mr. Nicholas that he was planning to move his family to Fort Wayne and get into the harness in the laboratory again on a full-time basis.

The problem of the projection tube was foremost in his mind. While there were many details which needed to be attended to in Newton Center and Fryeburg before he could move to Fort Wayne, he managed to spend the major portion of his time with the engineers working on the tube.

When I arrived in Fort Wayne for the directors' meeting at the end of May in 1948 I was delighted to meet Phil at the plant and learn from him that while they had not found a house, they planned to take up residence there and Pem was searching the city for a place. Phil was wholly immersed in the problem of the projection tube. Following the directors' meeting he took us all down into the laboratory and showed us how the new screen was fabricated and impregnated with the essential metallic materials. He, Chris Larson, and other members of the staff had built a laboratory model of the tube, and while they were not prepared to show us a picture on it, they wanted us to see the brilliant spot the surface would give. This demonstration was in a room brilliant with light from a big window. Everything was in readiness when we arrived and the spot was turned on. It was of blinding brilliance. Then Phil with his usual conciseness and clearness of exposition showed us how the screen was made, what the problems ahead were, and how they proposed to lick them. It was indeed a demonstration.

Pem and Phil finally found a house but could not get occupancy until fall. In the meantime they took an apartment and moved in in June. Progress on the tube has gone forward to the point where experimental pictures have been shown. In this development I was reminded of the development of the dissector tube and the seemingly endless work before this instrument, which is now the heart of the Farnsworth television system, became a reality.

In each succeeding time that I have seen Phil, he has given me reports on the projection tube for the receiver sets. In our last discussion Chris Larson joined in with the prediction that the tube had reached the stage where they felt that the end product would be satisfactory. How long it will take to bring the finished product to the commercial stage is still uncertain.

Coincidental with the development of the tube, Phil has a brilliantly conceived optical system to be used with it. There is good prospect that this latest Farnsworth invention will completely change the type of television receiver sets and make obsolete the large cathode-ray tubes which are a major handicap in getting a large picture without increasing the size of television sets to proportions which cannot be used in the ordinary living room.

The last time we were all together in Fort Wayne, Phil, Pem, Jess, and I had breakfast together. It was like old times. We took a long time because we had much to talk about. Principally, we talked about the launching of commercial television, which is now spreading rapidly in the larger cities. Thousands of new television receivers are being installed in homes every month, and the number of stations in operation will pass the hundred mark in 1949. Soon the nation will be joined together with television networks from coast to coast and from border to border.

The day before we left Fort Wayne we saw the handsome array of postwar television receiving sets which were to come off the Farnsworth production lines. There were table models, as well as models for sets in combination with radios and phonograph-radios to meet the pocketbooks of varying incomes. As we talked over what we had seen the day before, the four of us knew that at long last—after nearly a quarter of a century of continuous effort—the dream of television which Phil had had as a youth and in which Jess and I had believed, had finally come into its own.

TELECOMMUNICATIONS

An Arno Press Collection

Abramson, Albert. **Electronic Motion Pictures:** A History of the Television Camera. 1955

[Bell, Alexander Graham]. **The Bell Telephone:** The Deposition of Alexander Graham Bell in the Suit Brought By the United States to Annul the Bell Patents. 1908

Bennett, A. R. **The Telephone Systems of the Continent of Europe** and Webb, Herbert Laws, **The Development of the Telephone in Europe.** 1895/1910. Two vols. in one

Blake, George G. **History of Radio Telegraphy and Telephony.** 1928

Bright, Charles. **Submarine Telegraphs:** Their History, Construction and Working. 1898

Brown, J. Willard. **The Signal Corps U. S. A. in the War of the Rebellion.** With an Introduction by Paul J. Scheips. 1896

Chief Signal Officer, U. S. Signal Corps. **Report of the Chief Signal Officer: 1919.** 1920

Danielian, N[oobar] R. **A. T. & T.:** The Story of Industrial Conquest. 1939

Du Moncel, Count [Theodore A. L.] **The Telephone, the Microphone, and the Phonograph.** 1879

Eckhardt, George H. **Electronic Television.** 1936

Eoyang, Thomas T. **An Economic Study of the Radio Industry in the United States of America.** 1936

Everson, George. **The Story of Television:** The Life of Philo T. Farnsworth. 1949

Eyewitness to Early American Telegraphy. 1974

Fahie, J[ohn] J. **A History of Electric Telegraphy to the Year 1837.** 1884

Federal Communications Commission. **Investigation of the Telephone Industry in the United States.** 1939

Federal Communications Commission. **Public Service Responsibility of Broadcast Licensees.** 1946

Federal Trade Commission. **Report of the Federal Trade Commission on the Radio Industry.** 1924

Fessenden, Helen M. **Fessenden:** Builder of Tomorrows. 1940

Hancock, Harry E. **Wireless at Sea:** The First Fifty Years. 1950

Hawks, Ellison. **Pioneers of Wireless.** 1927

Herring, James M. and Gerald C. Gross. **Telecommunications:** Economics and Regulations. 1936

Lodge, Oliver J. **Signalling Through Space Without Wires:** Being a Description of the Work of Hertz and His Successors. [1900]

McNicol, Donald. **Radio's Conquest of Space:** The Experimental Rise in Radio Communication. 1946

Plum, William R[attle]. **The Military Telegraph During the Civil War in the United States.** With an Introduction by Paul J. Scheips. Two vols. 1882

Prime, Samuel Irenaeus. **The Life of Samuel F. B. Morse,** L.L. D., Inventor of the Electro-Magnetic Recording Telegraph. 1875

The Radio Industry: The Story of Its Development. By Leaders of the Radio Industry. 1928

Reid, James D. **The Telegraph in America:** Its Founders, Promoters and Noted Men. 1879

Rhodes, Frederick Leland. **Beginnings of Telephony.** 1929

Smith, Willoughby. **The Rise and Extension of Submarine Telegraphy.** 1891

Special Reports on American Broadcasting: 1932-1947. 1974

Thompson, Silvanus P., **Philipp Reis:** Inventor of the Telephone; A Biographical Sketch. 1883

Tiltman, Ronald F., **Baird of Television:** The Life Story of John Logie Baird. 1933

Wile, Frederic William. **Emile Berliner:** Maker of the Microphone. 1926

Woods, David L., **A History of Tactical Communication Techniques.** 1965